Mobile Phone Security and Forensics

Iosif I. Androulidakis

Mobile Phone Security and Forensics

A Practical Approach

Second Edition

Springer

Iosif I. Androulidakis
Pedini Ioannina
Greece

ISBN 978-3-319-29741-5   ISBN 978-3-319-29742-2   (eBook)
DOI 10.1007/978-3-319-29742-2

Library of Congress Control Number: 2016931614

Printed on acid-free paper

This Springer imprint is published by Springer Nature
The registered company is Springer International Publishing AG Switzerland

To my parents

Preface

Welcome to the second edition of "Mobile Phone Security and Forensics." The dominance of mobile phones has continued since the publication of the first version of the book, in an ever-increasing rate. However, while we are enjoying the technological advances that mobile phones offer, we are also facing new security risks coming as a cost of our increasing dependence on the benefits of wireless communications.

The purpose of this book is the same as before: to raise user awareness in regard to security and privacy threats present in the use of mobile phones. It is focused on practical issues and easy to follow examples, skipping theoretical analysis of algorithms and standards. Most sections have been enriched with new material. The book is more geared toward the mobile devices themselves and not the underlying networks, so most of the contents are applicable irrespectively of the "generation" of the network (GSM, 3G, 4G, etc.) to GSM and UMTS alike. The goal is to achieve a balance, including both technical and nontechnical chapters. Amateurs as well as experienced users will benefit from the overview of threats and the valuable practical advice. They will also get to know various tricks affecting the security of their phone. More advanced users will appreciate the technical discussions and will possibly try experimenting with the forensics and mobile phone control techniques presented in the respective chapters.

Chapter 1 gives an introduction to confidentiality, integrity, and availability threats in mobile telephones, providing the background for the rest of the book. In Chap. 2, the results of a large-scale survey and some following ones are presented, placing the user as one of the weakest links in the security landscape. With eavesdropping being one of the most apparent threats, a specific interception technique is examined in Chap. 3, while at the same time the inefficiencies of mobile phones' graphical user interfaces are highlighted in regard to security. The chapter is further enriched since the previous edition with a discussion regarding software defined radio and other advances in mobile telephony communications interception. Chapter 4 is the more diverse themed chapter of the book covering device and network codes, commands to control the phone, and software and hardware tricks. Software and mobile applications' security are not extensively covered since they mostly fall in computer security literature. Chapter 5 is devoted to security in SMS, as a leading service in mobile telephony. Moreover, there is an extended discussion for fighting unsolicited SMS messages (spam). Following, a chapter focusing on the procedures and techniques of forensics reminds us that mobile phones will sooner or later be criminals' preferred target. Concluding the book, Chap. 7 synopsizes the previous chapters and provides a condensed list of practical security advices users should follow.

Closing, I would like to thank my family for all the support and love, my professors in Greece and Slovenia for their mentoring during my studies, and the security researchers all over the world I have met and collaborated with. They are all too many to be listed here but they know who they are! Last, but not least, I would like to thank my editor and all of the members of the Springer team that I collaborated with. With such a good collaboration writing, the second version of the book was a true pleasure. I hope you will enjoy reading it. Writing a book is a hard and long process but thanks to my editor's guidance everything proceeded pleasantly and smoothly.

Ioannina, Greece   Iosif I. Androulidakis, Ph.D., Ph.D.
October 2015

Contents

Chapter 1
Introduction: Confidentiality, Integrity, and Availability Threats in Mobile Phones

Abstract In this introductory chapter, we will briefly describe and group as many as possible of the threats in Confidentiality, Integrity, and Availability that mobile phones are facing. As such, it will be the basis for the discussions that will follow in the next chapters.

Keywords Mobile phone threats • Mobile phone security • Mobile phone confidentiality • Mobile phone integrity • Mobile phone availability • GSM security • SMS security • Mobile phone interception • Mobile phone eavesdropping

1.1 Introduction

As it is known, GSM and next-generation technologies such as UMTS are digital systems for mobile, wireless voice and data communication that provide international access, high capacity, and transmission quality. Because of their many advantages over the older technologies, they conquered the world map of mobile communications, in a short period of time. Among their major advantages were voice and data encryption and security mechanisms combating fraud, a really positive development since analog systems were particularly vulnerable to both problems.

Unluckily for GSM, soon it was revealed that the provided protection was not enough and it was a matter of time and further research to bypass the corresponding security mechanisms and algorithms. The academic community and the applied practice have since proven that GSM suffers from a series of problems that negatively influence the Confidentiality, Integrity, and Availability of communications. While some of the problems have been solved in next-generation networks such as UMTS, there are still many left, especially the ones that are targeting devices, implementations, and services, rather than the protocols themselves [1–15].

For many of the weaknesses of the air interface, an actual exploit has not been easy to implement, since it evolved expensive and hard to find equipment. Recent research with open source software and hardware, however, has changed the landscape.

Apart from technical issues, the user is one of the weakest links, as the results of a relevant survey presented in Chap. 2 show. This is why this contribution mainly aims at raising user awareness. In this introductory chapter we will briefly describe

I.I. Androulidakis, *Mobile Phone Security and Forensics*,
DOI 10.1007/978-3-319-29742-2_1

and group as many as possible of the threats in Confidentiality, Integrity and Availability that mobile phones are facing, providing the background for the discussions that will follow in the next chapters.

1.2 Confidentiality

Starting with Confidentiality, the first threat that possibly comes to mind is eavesdropping and interception, not only of voice calls but also of data such as short messages or mobile web browsing. In addition, the rich multimedia content stored in modern mobile phones (photographs-video-sound recordings) as well as all kinds of data files including the contact list and the calendar/appointments list can get in the wrong hands. The potential use of the mobile phone for monitoring and spying the surrounding area (sound and video eavesdropping) is particularly worrying. Indeed, there are at least two ways this can be achieved: a phone with special software installed can monitor the surroundings while the phone appears to be in idle state, or, more simply, a small handset can stealthily be concealed in a meeting room and transmit the discussions to a listening post that could be in the other end of the globe. The information retrieved with such methods is of extremely high value, especially in cases of corporate espionage. More details will be presented in Chap. 4.

Interceptions can take place in an active and/or passive way. In the active method, there is some kind of physical or technical interaction with the target phone (or network) and as such it is easier to be spotted. Active ways of interception include Bluetooth attacks, malicious software that can be downloaded to the phone, launching a man in the middle attack posing as a legitimate part of the network (using a fake base station), and, of course, simple… theft! The man in the middle attack will further be explained in Chap. 3.

Passive interceptions, on the other hand, are possible by just "listening" to the airwaves and decrypting the encrypted radio wave communications that flow in the air. They are focused on cryptanalysis and can target protocols and algorithms in various parts of the network. In a typical wireless communication, the radio waves are transmitted freely in the air and they cannot be easily confined. Therefore, a potential intruder can intercept and process these signals without even coming close to the target. The most interesting parts to monitor are the mobile phone to base station communication (the radio interface of the mobile phone) and the base station to the rest of the network communication (usually achieved with microwave links, Fig. 1.1). More details on interception are found in [16–23].

Up to recently, it has been traditionally difficult to intercept air waves for GSM signals and to open source projects during the few past years along with commercial or do-it-yourself software defined radio peripherals have brought this functionality interact with mobile phones transmitting on their radio layer. Some open source projects during the past few years have brought this functionality to the interested user with great success. They are Universal Software Radio Peripheral (USRP) [24], GnuRadio [25], AirProbe [26], OpenBTS [27], OpenBSC [28] and

Fig. 1.1 Typical microwave links

OsmocomBB [29]. Some of these projects will be further discussed in Chap. 3. Moreover, one of the most recent and successful software tools to crack the voice encryption algorithm A5 is the "A5/1 Cracking Project" [30]. The project uses a form of time-memory tradeoff known as rainbow tables technique. According to this technique, all the possible encrypted strings for all the possible keys on a given plaintext are pre-computed and stored in a table. After that, a single intercepted session containing the encrypted plaintext can reveal the original key used to encrypt it. Having the encryption key, the rest of the communication can then be decrypted. Similar work has been presented in [31] too, using FPGAs. It must be also noted that a single old phone of a specific brand and model family can be used to passively intercept GSM traffic [32], albeit limited to the specific phone's communication. It is based on debugging functionality that apparently has made it to the final product. The easiest way to enable and use this feature is by using Gammu software [33]. In the rest of the book, we will not cover any further details since the content is not aimed at cryptanalysis but rather at practical attacks and security. The interested reader can find more about cryptanalysis of A5 in references [34–38].

Lawful interception (LI) functionality abuse can prove to be one of the most effective spying options. Although LI it is a very sensitive system, that should be adequately protected, successful attacks have been seen [39, 40]. Each provider is obligated to offer mechanisms to monitor communications for the needs of authorities, following the necessary warrant. Employees with the proper authorization credentials or malicious hackers that manage to break in the system (with or without internal help) can intercept at will the calls and the messages of the targets they chose.

Another direct privacy concern is the ability to monitor the mobile phone's position (and as such the user's position). Location tracking is traditionally used by law enforcement and the process of locating the phone involves the operator's help (information from the movement of the mobile phone according to the base stations it registers to). Modern phones, however, incorporating GPS technology make things simpler and more accessible to attackers. Even if the phone does not have a GPS sensor, specific malware-software can relay the identity of the cell that the phone is currently logged on (each cell covers a specific geographic area) to an eavesdropper. Then, there exist public databases listing the specific areas the cells serve. This way, the attacker maps the cell identity with the actual place the phone was last "seen" by the network. Various geolocation techniques exist such as [41, 42]. Exploiting SS7 functionality, it is also possible to coarsely locate a phone [43]. In a closer area, having switched on and visible Bluetooth lets others spot a user using Bluetooth "beacons" in specific areas and logging the time the user crossed that point (Fig. 1.2). It is also possible to locate a phone in a given area by its radio frequency transmission, using a directional antenna and the method presented in [44] (Fig. 1.3). RRLP, Radio Resource Location Protocol, used in some smartphones can be targeted, revealing the location of the phone to the attacker [45]. Finally, there are some cases of mobile phones uploading their activity and location to the network operator without user's consent.

1.3 Integrity

Integrity issues are directly related to interception, where the integrity of the communication cannot be guaranteed. Tampering with the handset (using hardware or software) is also possible. There are, however, more instances of integrity problems,

```
Scanning ...
Fri Oct 14 09:41:42 PDT 2005
Scanning ...
Scanning ...
Scanning ...
Fri Oct 14 09:42:54 PDT 2005
Scanning ...
Scanning ...
       00:01:E3:20:3▮▮▮5 0001E320▮▮▮
Scanning ...
       00:01:E3:20:3▮▮▮5 0001E320▮▮▮
Scanning ...
Fri Oct 14 10:05:46 PDT 2005
Scanning ...
       08:00:1F:1▮▮▮2:70 n/a
Scanning ...
Fri Oct 14 10:32:42 PDT 2005
Scanning ...
       00:11:9F:7▮▮▮▮ < to kinhto mou>
Scanning ...
       00:11:9F:7▮▮▮▮8 < to kinhto mou>
Scanning ...
       00:11:9F:7▮▮▮▮3 < to kinhto mou>
Fri Oct 14 10:33:03 PDT 2005
Scanning ...
```

Fig. 1.2 Bluetooth scanning beacon revealing the time users cross a certain point, effectively logging their movement

in connection to economic fraud involving mobile phones. In the early phases of GSM, cloning of SIM cards was possible. It required physical possession of the SIM to be cloned and special software that could extract the subscriber's authentication key (Ki) to be written to another SIM card and as such effectively "clone" it. Needless to say that the communication bills from that cloned SIM would be charged to the initial owner since it would appear to the network as the user making the calls. Further enhancements in the authentication algorithms made this attack obsolete. Of course, it is still possible with insider's help to acquire the Ki of a subscriber. Apart from bills, the cloned SIM would inherit the caller id of the victim too, enabling the attacker to launch an attack against integrity, masquerading as another user. There are even services in the Internet where the user can place calls choosing whatever caller id he wants. Masquerading occurs not only in voice calls but also in SMS. The identity of the sender can be changed in order to make a malicious message appear legitimate or for spam purposes. The phenomenon will be analyzed in Chap. 5.

Fig 1.3 Directional antenna locating mobile phone transmissions (after [44])

Fraud plays a major role in threats targeting the integrity of mobile phone communication [46–51]. Attacks sometimes begin inside the operator's network by malicious administrators. Mobile telephony network management is implemented using an array of different systems and platforms. A separate system usually exists for the communication and the technical operations and management and a different one for the billing. Moreover, it is possible to have a different billing platform system for the postpaid subscribers and another one for the ones using prepaid SIM cards. This complexity allows for a usual technical fraud as follows: The administrator activates a postpaid connection in the network but does not register the data in the billing platform. This mobile phone can make and receive calls without being accounted for.

One more "solution" exists for those with many bills unpaid, having their outgoing calling capability revoked due to debt. With some "internal" help, a direct tweaking in the database of the communication system can disable this barring. The billing platform (which usually is a different platform, as discussed) is still under the impression that connection is still barred and thus does not proceed to charging the traffic.

Segregation of postpaid and prepaid platforms allows for yet another technical internal fraud. It is possible to change the "profile" of a subscriber moving her from the postpaid connection to the prepaid platform, again without registering the change to the billing system. The prepaid billing system can't "see" the subscriber while the postpaid billing system does not "see" any traffic from the subscriber since traffic is now handled by the prepaid system.

Speaking of mobile operators, they are able to remotely send data and code to each SIM card (and mobile phone device) belonging to their network, irrespectively of whether the mobile phone has internet connectivity. This is possible with technologies such as OTA (over the air) and binary SMS [52]. In Chap. 5, we will examine some more technical details. An obvious case for such functionality would be to have automatic firmware upgrades (FOTA — firmware over the air) of the mobile phones as well as new installation of updates and new versions of programs, or automatic settings configuration (e.g. OMA-DM technology — Open Mobile Alliance Device Management). Such techniques make it unnecessary to visit customer support centers and the user has a better experience. It is clear, however, that in the hands of wrong persons it can prove to be a very powerful weapon, tampering with the integrity of the mobile phone for interception or fraud (e.g. by installing software without users' permission).

1.4 Availability

Given mobile phone's omnipresence, its availability has been considered a given fact. Truth is, however, that it is very easy to block mobile phone communications [53, 54]. Denial of service attacks can be escalated using RF (radiofrequency) jamming. The attacker jams the radio spectrum in the band of operation of the mobile phones transmitting high-power electromagnetic noise as we will see in Chap. 4. Depending on the power of the transmitting jammer, the area of denial of service can extend from one room to a whole city. Dozens of such products are available in the internet, although their legitimacy is questioned. Network availability can also be stressed and ultimately filled by using special transmissions in the radio interface of GSM. This is possible using some specific mobile phones. The firmware of these phones has been leaked and as such they can reprogrammed to act in ways not foreseen by the standard. In a typical technique, the attacker sends multiple access requests continuously to the network, without waiting for the timeout, eventually filling up all the signaling channels. This way, no other mobile phones in the area can initiate a call [29]. The other way around, the base station can notify a mobile phone that should stop operating, possible due to being stolen, as presented in [15].

A different (and possibly more annoying) way of attacking the availability of a phone is to place consecutive calls to it, effectively making it ring nonstop. The user is then forced to switch it off. With more sophisticated techniques, targeting the network backbone and its signaling, an area or an entire city can become the target of the attackers, banning phone calls altogether.

SMS are also particularly interesting for denial of service attacks, both to the network infrastructure and also to a specific mobile phone. Flooding the network (or the handset) with SMSs leads to buffers' exhaustion causing various reactions, ranging from loss of messages to crashes and even (semi)permanent blocking of the mobile phone. The latter is also easily achieved by using specially crafted binary messages or specific combinations of SMS parameters [55–58]. When such a message arrives, the phone does not know what to do with it and reacts in a nonpredicted way. In any case, battery exhaustion, using either consecutive calls or sending silent SMSs is by far the most effective way to go, since without power the mobile phone is pretty much useless!

Research shown in [59] has shown it is very easy to cause a selective denial of service attack targeting the mobile phone of the victim, with minimal cost and resources. Under specific conditions, an attack of 2 h only was sufficient to completely deplete the battery, leading to mobile phone service denial of service. The consequences of such attacks can vary, from simple annoyance to lost business and even life-threatening instances in life and death situations where the user is deprived from mobile phone service. In any case, they are completely stealth without the victims being able to realize they are under attack. Stealth attack scenarios are based on the repetitive reception of stealth SMSs and/or very short calls that get interrupted before the mobile phone rings, and as such do not show up at all. These attacks force the mobile phone to keep transmitting and consume energy, leading to battery depletion in a short time.

For the SMS sending attack vector, 3GPP 23.040 [60] specifies a special type of SMS with protocol-ID 64 (TP-PID Short message type 0) that resembles the "ping" network command of computer operating systems. It is unconditionally received (even when memory is already full) and automatically deleted from the mobile phone without the user ever realizing. Since providers prefer to disable this kind of SMS in their networks, a better method as detailed in [61] involves using another type of "invisible" SMS. This specific type of SMS is used in order to switch on or off the voice message, email, or "other type of message" indicators mobile phones have. Such SMSs operate according to the data coding scheme of the SMS TP-DCS (3GPP 23.038 [62]) or using the User Data (TP-UD) part of SMS [60].

The attacker sends an SMS that switches off the "other type of message" waiting indicator (via TP-DCS = 195). This waiting indicator is never used in mobile phones, so it is not expected to ever be lit. By sending a command to the phone to switch off an icon that is already switched off, no visible action is produced and the victim is not aware of the process taking place. Interestingly, the screen of the mobile phone remains dark during the reception and the processing of this message, enforcing the stealth character of the attack. The mobile phone, however, is forced to transmit, and therefore consume energy, according to the will of the attacker. As a side note, such a "ping" SMS can also reveal areas where the user's mobile phone is left without network coverage (such as in an elevator or in a tunnel). This way, an attacker can possibly know the exact moments the target is crossing some specific landmarks in a daily journey [61], accordingly timing his denial of service attack.

The special SMSs needed can be sent either by using a GSM modem, a mobile phone connected to a PC, or, even easier, by using Internet-based SMS services, such as bulk SMS providers offer. Given the low cost of SMSs, many thousands of them can be sent with a minimal cost, for the whole duration of the denial of service attack. The cost can be even zero, if a prepaid SIM card is used (that usually allow free SMSs among users of the same provider). It is also possible to use a professional GSM tester, as described in [22], but again for shake of simplicity it might be easier to work with a bulk SMS provider. In addition, the attack using the GSM tester would not be stealthy enough. The SMSs would be still invisible, but during the attack using the GSM tester, the target is "stolen" from the originally serving network and as such cannot receive any incoming legitimate traffic.

Continuing the analysis, the repeated very short calls scenario can be carried out by programming a Private Branch Exchange (PBX) with Integrated Services Digital Network (ISDN) connectivity to call the mobile phone but hang the line up before the mobile phone actually rings. Instead of a PBX, an ISDN modem can also be used to realize controlled duration calls from a PC. An ISDN line is essential since it offers very fast signaling which is again the core idea. With an ISDN line, call duration can be precisely tailored, achieving a resolution of 100 ms or even better. Quite interestingly, due to the way the GSM radio interface operates, a call can take up to 8 s of signaling before the mobile station actually rings [63], according to the paging process described in [64]. Initially, a downlink transmission from Base Transceiver Station (BTS) signals the incoming call to the paged Mobile Station (MS) using the common Paging channel (PCH). Following, the MS (the GSM terminal) demands access to the network via the Random Access Channel (RACH). The BTS replies using Access Grant Channel (AGCH) that further establishes a Standalone Dedicated Control Channel (SDCCH) to cater for the bidirectional signaling. Finally, a traffic channel (TCH) gets reserved to host the conversation for the call to continue with the actual voice content (Fig. 1.4).

If the originator hangs up before the completion of this cycle, then the mobile phone will not ring. Nonetheless, it will have consumed energy in order to perform the initial steps of signaling, request the dedicated channel, and so on up to the point where the call was terminated. During the tests reported in [59], using short calls of 1000 ms (milliseconds) worked perfectly well under all situations producing signaling duration of 1500 ms. It must be noted that if the mobile phone is already engaged in a call then a dedicated signaling channel is already established and, thus, the call

Fig. 1.4 Signalling
between MS and BTS

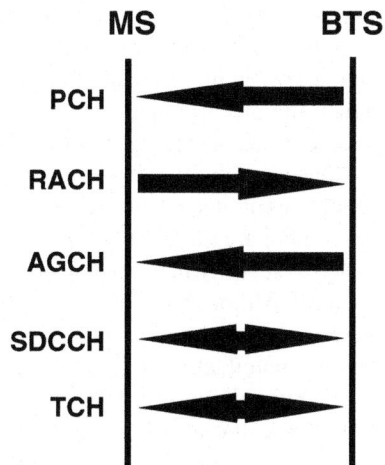

will show up in the display of the phone if it lasts more than 300 ms. Consequently, if the attacker wants to be completely sure that will not be revealed, he should set the short calls duration to less than 300 ms. Another limitation that must be addressed is the fact that if the mobile phone is switched off before or during the attack and the user had opted to be informed about missed calls, then the short call is again registered if it lasts for more than 300 ms. Indeed, now the network knows that the subscriber has the phone switched off it immediately registers the call since there is no need to page the phone. Therefore, the duration of the signaling process is far shorter. In order to cater for this, the attacker can use the stealth SMS types, in the same way it was described earlier, to be informed about whether the mobile phone of the target is switched on or off.

Let's not forget that interruptions in the service are not always caused by attackers. More than often, technical glitches, bugs, or environmental disasters cause extended scale and duration service interruption incidents.

Another factor that often slips our attention is the probability of the mobile phone's theft. With all these advanced features that have closed the gap between computers and telephones, more information (including sensitive) is stored in the mobile phones. Being small items, handsets can be misplaced or stolen. As a matter of fact, the theft can be only temporary, lasting a few minutes. It is only that long it takes for a malicious person to install software or even electronic circuits to the device, turning it to an interception device spying its owner.

1.5 Manufacturers' Responsibilities

Apart from users, telecom providers, and their employees, responsibilities for the security also lay with mobile phone devices manufacturers. The rate of promotion and marketing of new devices, operations, and services in the market is overwhelming. At the same time, there is extensive complexity and significant systems interaction. As such, thorough testing of all possible scenarios and parameters that can lead to vulnerabilities is a process that cannot be applied with absolute success. This technical difficulty in combination with the stringent time margins and deadlines to launch the products and services makes things worse. The result is of course gaps and inefficiencies in regard to security in all aspects.

Consequently, it is a matter of time before vulnerabilities are discovered (and possibly exploited), from researchers in the best case or from criminals in the worst case. There are quite many examples where bad implementations from manufacturers led to security incidents. It was also the case with previous versions of Bluetooth standard. Although its security stack was carefully enough designed in the standard, there were many wrong and sloppy implementations in handsets themselves from various manufacturers resulting in many vulnerabilities and relative incidents. It must be noted that social engineering, the technique to convince users to do something while the attacker is possibly masquerading as somebody else (colleague, boss, technician etc) was also an important parameter in the success of these attacks.

A last comment in regard to manufacturers' responsibilities is relevant to security interfaces. In Chap. 3 we will specifically examine a specific GUI (Graphical User Interface) shortcoming in regard to security. While in early mobile phones the GUI security functionality was indeed limited because of the small and monochrome screen, nowadays, manufacturers shipping mobile phones with huge screens have no excuse not to implement security indicators and features in a usable and fully informative context that can protect the user from the threats she is facing.

1.6 Malicious Software and Other Issues

Bluetooth was for many years the main vector of mobile phone virus spreading, targeting the first generations of advanced phones. Mobile phones evolved and became "smart" phones, being closer to personal computers and able to do almost everything a PC can do, including Internet browsing. In modern mobile phone operating systems, developers can write code the same way they do with computers and there are literally hundreds of thousands of applications already available. However, besides the increased usability, the more complex a system is the more security gaps it has. Along with legitimate applications, viruses and other malicious programs can target the mobile phones [65–78].

Respectively, dozens of interception programs are available that when installed in a given mobile phone they broadcast in detail to the attacker the behavior of the phone (and its user) [79]. More details will be presented in Chap. 4.

Closing, Internet access in mobile phones, along with its many goods, opens the door to all kinds of traditional threats computers face. As a matter of fact, it seems that mobile phones will be the platform of choice for cyber criminals of the future.

1.7 Conclusion

In conclusion, Confidentiality, Integrity, and Availability of communications in mobile phones face threats of technical and non-technical nature, in many different levels. As we will shortly see in the next chapter, users are not adequately informed in regards to the threats they face, and as such they do not follow the best practices to protect their mobile phones. They are also not familiar with all of the functions the mobile phone offers, while graphical user interfaces do not always help promote security. Therefore, the problem is not only technical but it also extends to users' education and awareness as well as human–computer interaction via the graphical user interface.

To combat the threats, before rushing to technological solutions, the first step could be the awareness and better education of users. Following this introduction, more details and practical security issues will be highlighted in the next chapters, hopefully raising readers' awareness level.

References

1. Suominen M. GSM Security. Helsinki University of Technology
2. Steve Lord, Modern GSM Insecurities, 2003 . GSM-Security.net
3. Huynh T, Nguyen H. Overview of GSM and GSM Security, Department of Electrical Engineering and Computer Science Oregon State University
4. Quirke J (2004) Security in the GSM system., AusMobile
5. Gadaix E (2001) GSM and 3G Security
6. Gadaix E (2006) NGN Security. Bellua Cyber Security 2006
7. Gadaix E (2003) GSM operators security, xcon
8. Preneel B. Mobile network security. Katholieke Universiteit Leuven
9. Lord S (2003) Trouble at the Telco: when GSM goes bad. Network Security 1:10–12
10. Yousef P. GSM-Security: a Survey and Evaluation of the Current Situation. ISY, Linköping Institute of Technology
11. Androulidakis I (2009) Security in GSM and in mobile phones. IT Security Professional Magazine 9:35–41
12. Androulidakis I (2006) This is how hackers hack into our cell phones. Sunday Newspaper "To proto thema". 90:40–41
13. Androulidakis I (2006) Security issues in cell phones. Defence and Diplomacy Magazine 187:100–102
14. Karsten N, Krißler S (2009) Subverting the security base of GSM, HAR2009
15. Karsten Nohl and Chris Paget, GSM—srsly? 26C3, Berlin 2009
16. Pesonen L (1999) GSM interception. Department of Computer Science and Engineering Helsinki, University of Technology
17. Shoghi Communications Limited. Interception and monitoring of SMS & Voice communications on GSM 850/900/1800/1900 MHz Networks
18. Cryptome.org. Interception of GSM Cellphones, 2005
19. Patel S (2004) Eavesdropping without breaking the GSM encryption algorithm. 3GPP TSG SA WG3 Security—SA3#33, 10-14 May, Beijing
20. Manuel J (2002) Fernandez-Iglesias on the application of formal description techniques to the design of interception systems for GSM mobile terminals. J Syst Softw 60:51–58
21. Androulidakis I (2009) Intercepting mobile phones. IT Security Professional Magazine 8:42–48
22. Androulidakis I (2011) Intercepting mobile phone calls and short messages using a GSM tester, vol 160, Springer communications in computer and information science. Springer, Berlin, pp 281–288
23. Rieger F (2005) New interception threats from non-state actors and software-based voice encryption. IEE Secure Mobile Communications
24. ETTUS USRP, www.ettus.com
25. GnuRadio, http://gnuradio.org
26. AirProbe, http://svn.berlin.ccc.de/projects/airprobe
27. OpenBTS, http://openbts.sourceforge.net
28. OpenBSC, http://openbsc.osmocom.org/trac/wiki/OpenBSC
29. OsmocomBB, http://bb.osmocomm.org
30. The A5 Cracking Project, http://opensource.srlabs.de/projects/a51-decrypt
31. Hulton D, Mueller S (2008) Intercepting Mobile Phone/GSM Traffic. BlackHat Europe 2008
32. Tracelog, http://svn.berlin.ccc.de/projects/airprobe/wiki/tracelog
33. Gammu, http://www.gammu.org
34. GSM A5 files on Cryptome, http://cryptome.org/0001/gsm-a5-files.htm
35. Biryukov A, Shamir A, Wagner D (2000) Real time cryptanalysis of A5/1 on a PC. In: Proceedings of fast software encryption, New York, Lecture Notes in Computer Science. Springer, Berlin
36. Barkan E, Biham E, Keller N (2003) Instant Ciphertext-Only cryptanalysis of GSM encrypted communications. In: Boneh D (ed) CRYPTO 2003, vol 2729, LNCS. Springer, Heidelberg

37. Golic J (1997) Cryptanalysis of alleged A5 stream cipher. http://cryptome.org/jya/a5-hack.htm
38. Briceno M, Goldberg I, Wagner D. A pedagogical implementation of the gsm A5/1 and A5/2 voice privacy encryption algorithms. http://www.cryptome.org/gsm-a512.htm
39. Vodafone Griechenland im Visier der Ermittler, dsltarife.net/news, 2006
40. Prevelakis V (2007) The Athens affair. IEEE Spectrum, July
41. Laitinen H, Lahteenmaki J, Nordstrom T (2001) Cellular location technology, 2001. Conference: Vehicular Technology Conference
42. Warnock M. Geolocation via cell tower data
43. Engel T (2008) Locating mobile phones using signaling system #7, 25th Chaos Communication Congress
44. Androulidakis I (2011) Locating a GSM phone in a given area without user consent. Presentation in hack.lu 2011 conference, Luxembourg, 19 Sept 2011
45. Welte H (2009) Report of OpenBSC GSM field test, HAR2009
46. Hynninen J (2000) Experiences in mobile phone fraud. HUT TML
47. Müller M (1999) Intruder scenarios in telecom networks
48. Shawe-Taylor J, Howker K (1999) Detection of fraud in mobile telecommunications. Information Security Technical Report 4(1)
49. Androulidakis I (2011) Combating telecommunications cybercrime, 3 hour course, high-tech crime department of the national bureau of investigation of Hungary, Budapest, 19 April 2011
50. Androulidakis I. Detecting cybercrime in modern telecommunication systems. In: European Police College (CEPOL), Seminar 64/2010, Cyber Crime & High Tech, Athens, 18–21/5/2010
51. Androulidakis I. Cybercrime in mobile telephony systems, European Police College (CEPOL), Seminar 62/2011, High Tech & Cyber Crime, Brdo near Kranj, Slovenia, 20 Oct 2011
52. Cadonau J (2008) OTA and secure SIM lifecycle management smart cards, tokens, security and applications
53. Bocan V, Cretu V (2006) Mitigating denial of service threats in GSM networks, ARES
54. Bocan V, Cretu V (2004) Security and denial of service threats in GSM networks. Trans Automatic Control Comput Sci 49(63)
55. Miller C, Mulliner C (2009) Fuzzing the phone in your phone. http://www.blackhat.com/presentations/bh-usa-09/MILLER/BHUSA09-Miller-FuzzingPhone-SLIDES.pdf
56. Mulliner C, Golde N, Seifert J-P (2011) SMS of death: from analyzing to attacking mobile phones on a large scale, 20th USENIX Security Symposium
57. Windows phone SMS attack discovered reboots device and disables messaging hub. http://www.winrumors.com/windows-phone-sms-attack-discovered-reboots-device-and-disables-messaging-hub, 2011
58. Engel T (2008) Remote SMS/MMS denial of service—"curse of silence" for Nokia S60 phones. http://berlin.ccc.de/~tobias/cursesms.txt
59. Androulidakis I, Vlachos V, Chatzimisios P (2015) A methodology for testing battery deprivation denial of service attacks in mobile phones. In: International Conference on Information and Digital Technologies (IDT), 7-9 July 2015, pp 6–10. doi:10.1109/DT.2015.7222942
60. 3GPP, 3rd Generation Partnership Project, Technical Specification Group Core Network and Terminals, Technical realization of the Short Message Service (SMS), 3GPP TS 23.040, 2010.
61. Androulidakis I, Basios C (2008). A plain type of mobile attack: Compromise of user's privacy through a simple implementation method. In: 3rd International conference on communication systems software and middleware and workshops, , 6–10 Jan 2008. COMSWARE 2008, pp 465–470doi:10.1109/COMSWA.2008.4554458
62. 3GPP, 3rd Generation Partnership Project, Technical Specification Group Core Network and Terminals, Alphabets and language-specific information, 3GPP TS 23.038, 2010.
63. Kune DF, Koelndorfer J, Hopper N, Kim Y (2012) Location Leaks on the GSM Air Interface, Internet Society. In: 19th Annual network & distributed system security symposium, ISOC-NDSS
64. 3GPP, 3rd Generation Partnership Project, Technical Specification Group Core Network and Terminals, Mobile radio interface Layer 3 specification, Core network protocols, 3GPP TS 24.008, 2015

65. Morreeuw J (2002) Securite des mobiles GSM
66. Hypponen M (2005) Mobile phone threats, HITBSecConf2005. Kuala Lumpur, Malaysia
67. Grand J (2004) Introduction to mobile device insecurity. Black Hat Europe
68. Sima C (2004) Security for handhelds and cell phones attacks and theories. Interoop Las Vegas
69. Greene K (2007) Securing cell phones. MIT Technology Review
70. Bickford J, O'Hare R, Baliga A, Ganapathy V, Liviu I (2010) Rootkits on smart phones: attacks, implications and opportunities, HotMobile'10, Annapolis, Maryland, February 22-23
71. Miller C, Honoroff J, Mason J (2007) Security evaluation of Apple's iPhone. Independent Security Evaluators
72. Mulliner C (2005) Exploiting PocketPC, what the hack
73. Mulliner C (2006) Security of smart phones. University of California, Berkeley
74. Mulliner C. Using labeling to prevent cross-service attacks against smart phones, DIMVA2006
75. Mulliner C (2008) Attacking NFC mobile phones. EUSecWest
76. Mulliner C (2006) Advanced attacks against PocketPC phones. DEFCON 14
77. de Haas J (2005) Symbian phone Security, Blackhat
78. Spaar D (2009) Playing with GSM RF interface, 26C3, Berlin
79. The Spyphone Guy. http://www.spyphoneguy.com/

Chapter 2
A Multinational Survey on Users' Practices, Perceptions, and Awareness Regarding Mobile Phone Security

Abstract In this chapter, we will present some interesting findings from a large-scale empirical study. It was conducted in a sample of 7172 students studying in 17 Universities of 10 European countries, in order to assess users' levels of security feeling and awareness in regard to mobile phone communications. As this study revealed, there are categories of users who face increased security risks due to their self-reassuring feeling that mobile phones are per se secure. These users feel that mobile phone communication is secure and tend to be less cautious in their security practices. There was also a statistically backed correlation of an array of demographics and usage characteristics and practices to the overall security level of the user. As such, specific profiles of users were extracted according to their mobile phone objective and subjective security level.

Keywords Empirical study • Security survey • User awareness • Security awareness • Security perceptions • Mobile phone security • Security practices • User profiling • Mobile downloading • Ciphering indicator

2.1 Introduction

Mobile phones have become a vital part of daily life for billions of people around the world. Their presence is ubiquitous and most users report that their cell phone makes them feel safer, even sleeping with their phone on or right next to their bed [1]. As described in the previous chapter, this dependence on mobile phones is not free from security risks. In this chapter, we will present some interesting findings from a large-scale empirical study. It was performed in order to assess users' levels of security feeling and awareness in regard to mobile phone communications and the results were published in various academic conferences [2–7].

As this study revealed, there are categories of users who face increased security risks due to their self-reassuring feeling that mobile phones are per se secure. These users feel that mobile phone communication is secure and tend to be less cautious in their security practices. Moreover, there was a statistically backed correlation of an array of demographics and usage characteristics and practices to the overall security level of the user. This way, specific profiles of users were extracted according to their mobile phone objective and subjective security level.

These categories of users need proper training and education; otherwise, a security incident will soon follow, harming in the long term the network operators too. They must be protected from unauthorized third party access to their data and from economic frauds. It is unquestionable that since users fail to secure their phones, reinforcing their security level should become a critical imperative.

2.2 Methodology

A very useful evaluation method for surveying user's practices is the use of multiple-choice questionnaires (i.e. in person delivery or e-mail questionnaires) [8, 9]. This empirical survey was conducted in the first half of 2010 using in-person (face to face) delivery technique, with a total of 7172 respondents participating (students in 17 Universities of 10 European countries).

This method was selected from other alternatives because it is more accurate and has a bigger degree of participation from the respondents (e-mail questionnaires usually are treated as spam mail from the respondents plus there is the risk of misunderstanding some questions). Indeed, the approximate ratio of participation was 80 % since the researchers were able to answer the questions of participants regarding the scope and the purpose of the survey. There was also a pilot study, conducted in the University of Ioannina, Greece, before the questionnaire was administered to the sample, to ensure the reliability and validity of the questionnaire [10]. At this point, it is interesting to note that there are not available already validated questionnaires for the subject.

The target group of the survey was university students from ages mostly 18–26, incorporating both younger and older youth segments because these ages are more receptive to new technologies. Given the fact that nowadays a very high percentage of young people is studying, the sample is not deemed limited and can be considered as representative of a large percentage of general youth population. Furthermore, since they are still studying, it would be easier to participate in security education programs, possibly implemented in Universities.

The English questionnaire was prepared, containing 22 questions, including the demographic ones. It was divided into two parts. In the first part, participants were asked demographic questions including gender, age, and field of studies as well as some economic data including mobile phone usage, connection type, and budget spent monthly on phone service. In the second part, specific questions related to their practices and security perceptions regarding mobile phones' security issues were researched. The questionnaire was translated to the corresponding languages. The translated text was reviewed by a third native language user to spot any translation errors or cultural misunderstandings. Data entry, finally, took place using custom software [11] while processing was done with SPSS.

Answers in specific questions were correlated to the answers in question: "Are you informed about how the options and technical characteristics of your mobile phone affect its security?" which had the following possible answers: "A Very Much, B Much, C Moderately, D Not too much, E Not at all". Apart from the statistical interpretations, a simple mathematical formula was also developed in the analysis of the security knowledge to produce numerical values from the multiple choice questionnaires. Responses were weighted with these weights: Very Much: 4, Much: 3, Moderately: 2, Not much: 1, Not at all: 0 and then divided by the number of occurrences, in order to get a mean value that was called "Mean Security Feeling Value (MSFV)".

In addition to MSFV which was based on subjective answers, another, objective, metric was introduced, the "Mean Actual Security Value (MASV)". MASV was calculated by adding one point for each of the following practices, which are objectively correct: Having IMEI noted down, knowledge of lack of encryption icon, having SIM PIN enabled, using a screensaver password, having Bluetooth disabled, not lending the phone, not downloading software to the phone, using antivirus, not saving passwords in the phone, and not saving personal data in the phone. The maximum score would hence be 10, since there were 10 specific questions.

Similarly, an objective awareness metric was introduced, the "Mean Actual Awareness Value-MAAV". For this one, one point was added for each "I do not know" in the answers. The maximum score would hence be 7 denoting a highly lacking awareness profile while 0 would be the mostly security aware score (negative scale).

2.3   Results

2.3.1   In General

In the next sections we present the results of categorizing users in regard to their security knowledge using the correlation and the simple formula described earlier. All of the findings presented are statistically significant at the Pearson's Chi-Square test $p < 0.001$ level. As was found, there are many statistically significant correlations between the following parameters:

- Security feeling and awareness to {sex, age, field of study, brand, operating system of phone, monthly bill}
- Country to {downloading, security awareness, feeling, practices}
- Storing personal data to {security awareness, feeling, practices}
- O/S type (advanced or not) to {downloading, security awareness, feeling, practices}
- Bluetooth usage to {demographics, security practices}
- Downloading to {demographics, security awareness, feeling, practices}.

These findings are thoroughly presented in papers [2–7].

2.3.2   Demographics

Among the 7172 participants of Table 2.1, 53 % were females and 47 % were males while most of the respondents were aged 18–26 (75 %). The subjects were studying various sciences and were generally equally distributed.

Regarding mobile phone usage, almost 67 % of them are using daily a single mobile phone, with some 24 % using two phones regularly. Nokia is the favorite brand, reaching 39 % of students followed by Sony-Ericsson (25 %) and Samsung (15 %). Apple's iPhone seems to be scarce among students with less than 4 % of penetration. It is immediately apparent that focusing on Nokia and Sony-Ericsson phones, a security awareness campaign would immediately target almost 2/3 of users yielding a very high return of investment. The brand itself however is not enough to categorize attack vectors and practices, since there is also the feature of the specific operating system running on each phone.

The Mean Security Feeling Value MSFV was 2.26 in the scale 0–4 (0 not at all, 4 very much), with minimal differences among genders. Correspondingly, the Mean Actual Security value MASV was calculated to be just 3.55 out of maximum value 10. The MSFV was found to be somewhat higher in younger ages. Examining the field of study, we discovered that soon to be medical doctors are feeling the most secure (MSFV 2.69). Mathematics and Natural Science students with MSFV 1.89 were in the other end of spectrum the most worried ones. Engineers were in the middle of the range, with MSFV 2.24.

2.3.3   Economics

Proceeding to economics, participants were asked whether they are using a prepaid or post-paid (contract) mobile phone connection. 42.4 % of students are using a contract-based subscription, a rather high percentage, while 13.6 % have both prepaid and post-paid SIMs (Subscriber Identity Module). Users having both types of connection seem to be more worried about security issues. Answering how much money they spent monthly, student mobile phone users had a wide range of financial capabilities. The leading 36.7 % spends 11–20€ (currency converted) monthly while 30.5 % spend less than 10€. Only 9 % spend 31–40€ and some 6.3 % spend more than 40€ per month.

The MSFV shows an interesting trend. It progressively gets lower as the bills get higher, from 2.33 (≤10€ bill) to 2.05 (31–40€ bill). Then, for users who spend more than 40€, it grows a little to 2.12. This is quite logical, since the more users spend, the more are concerned about the security of communication and possible fraud.

Following with a question of both security and economic importance, almost half of participants (47 %) do not download any software at all. There is also a 19 % that actively downloads ringtones or logos while some 16 % do not know whether their phone is able to download or not. The combined downloading mean including Ringtones/Logos, Games, and Applications is around 37 %. Of course, getting familiar with downloading users is being more vulnerable to downloading and using unauthorized software that can harm their phone.

Table 2.1 Sample distribution

	Country	City	Univ.	Students	University name		
1	GR	Ioannina	1	780	Univ. of Ioannina		
2	BG	Sofia	1	991	Univ. of Sofia		
3	RO	Iasi	3	994	Gheorghe Asachi Technical Univ.	Univ. of Medicine and Pharmacy "Gr. T. Poppa"	Alexandru Ioan Cuza Univ.,
4	CZ	Brno	2	633	Masaryk Univ.	Brno Univ. of Technologty	
5	SK	Bratislava	1	509	Comenius Univ.		
6	HU	Budapest	4	959	Semmelweis Unin.	Budapest Business School	Eotvos Lorand Univ.	Corvinus University
7	LT	Siauliai	1	759	Siauliai Univ.		
8	LV	Riga	2	620	Univ. of Latvia	Riga Technical University	
9	EE	Tallinn	1	829	Univ. of Tallinn		
10	SI	Ljubljana	1	98	Univ. of Ljubljana		
10		17		7172			

2.3.4   Security-Specific Questions

Our fundamental research question was how "secure" users feel that mobile phone communication is. The majority (36.9 %) replied "moderately" followed by 28.6 % "much" (Fig. 2.1). On the other hand, some 21.36 % felt not too much or not at all sure they are secure. Using the simple formula described in Sect. 2.2 the mean security feeling value (MSFV) was 2.26, in the 0–4 scale (0 not at all, 4 very much).

In addition, students answered whether they are informed about how the options and the technical characteristics of their mobile phones affect the security of the latter and whether they are taking the necessary measures to mitigate the risks. The majority (30.8 %) states that they are "moderately" informed while a large 15.8 % believes that they are "not at all" informed (Fig. 2.2).

Correlating MSFV value to awareness feeling (Fig. 2.3), an almost linear relationship between them manifested. Users who feel very much informed believe that communication is very much secure. On the other end, users who do not feel informed are afraid that communication is not at all secure. At this point, one can argue that excessive confidence can lead to "relaxation" of security practices. In addition, a campaign to enhance the security knowledge of users would lower their fear of communication insecurity, probably leading to greater phone usage and profits for the operators.

There was an even better (negative) linear association between the subjective security feeling and the objective mean actual security value (Fig. 2.4). Users who believe that mobile phone communication is very much secure have the lowest Mean Actual Security Value MASV (3.44). That shows a clear discrepancy between user opinions on security and actual security practices. The association grows linearly to the highest MASV of 3.84 for those that believe that communication is not secure at all. This group employs the most best practices, bit still fails in more than half (c.f. Methodology, where the maximum value of MASV is theoretically 10)

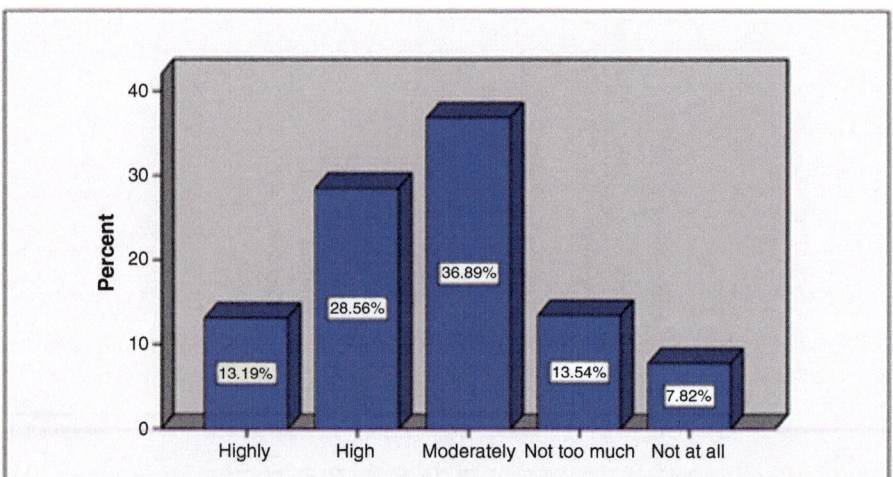

Fig. 2.1 How secure do you consider communication through mobile phones?

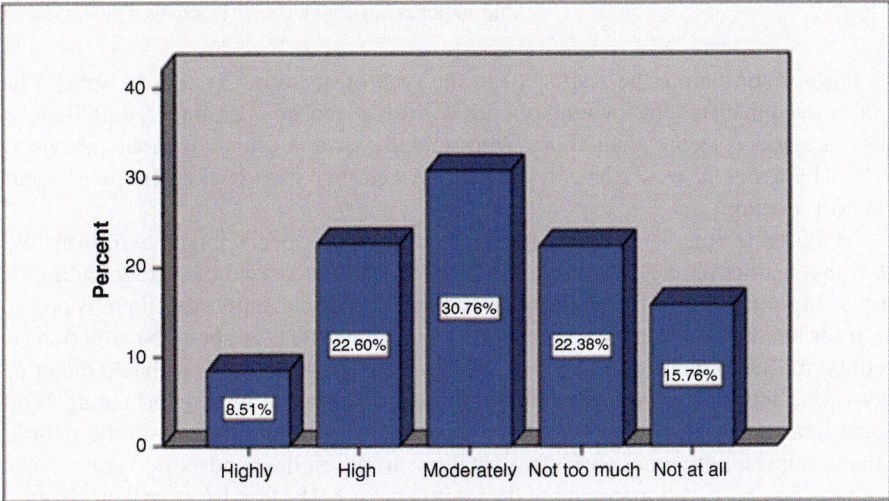

Fig. 2.2 Knowledge of mobile phone security aspects

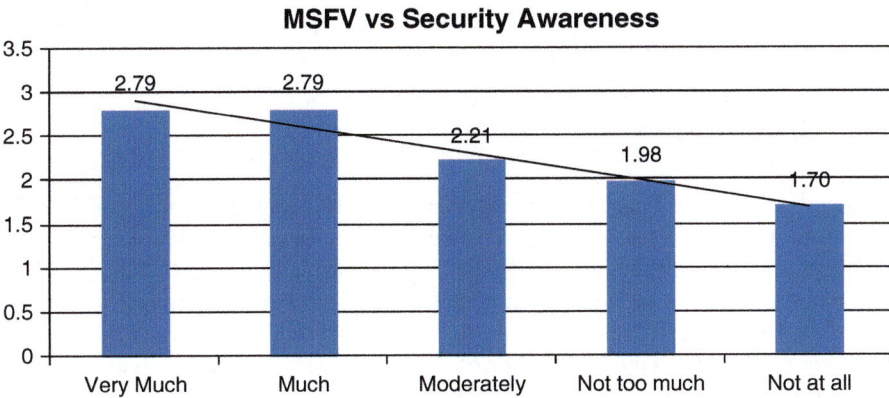

Fig. 2.3 Mean security feeling value vs. security feeling

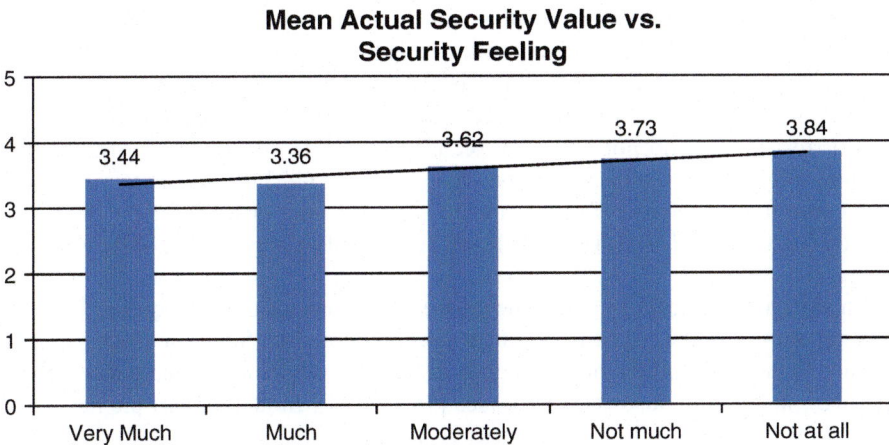

Fig. 2.4 Mean actual security value vs. security feeling

Further correlating the responses to the type of operating system (advanced or not) proved that students owning phones with an advanced operating system believe they are more secure than those who actually own a phone without advanced O/S. There was also a clear connection between increased backup frequency and security feeling.

At the same time, knowledge of the existence of the special icon that informs the user that his/her phone encryption has been disabled increased the safety feeling of users. In short, when A5 encryption is switched off or not supported, there is provision for handsets to display a special icon informing the user about the situation (it will be further explained in Chap. 3). Such an occurrence can be attributed either to network's lack of encryption capability or to temporary failure/overloading. The same icon can appear when a malicious attacker is launching a man in the middle attack, impersonating network's base stations to deceit the handset into connecting with the fake base station instead of the legitimate one. The fraudster can then channel the communication through his own equipment, effectively intercepting it [12]. This finding is a clear explanation of how better User Interfaces can help enhance the subjective security feeling via an objective method.

2.4 Related Work

Although there have been quite many theoretical studies concerning mobile services and mobile phones, a significant means for investigating and understanding users' preferences is asking their opinion via specific questioning techniques. The vast majority of these surveys indicate the growing importance of mobile phones in everyday life and the increased popularity of new features [1, 13]. In addition, with the apparent omnipresent availability of wireless devices, mobile services have a very promising prospect [13]. However, the success of those services (namely, m-commerce and m-payment) depends much on the security of the underlying mobile technologies [14], and mobile ubiquitous services pose great security challenges [15] Furthermore, users are interested in mobile services adoption only if the prices are low and the security framework is tight enough [16, 17].

A study of mobile users focusing on their awareness and concerns related to security threats, from security vendor McAfee, indicated that more than three quarters of respondents do not have any security at all [18]. In other words, despite acknowledging the wealth of threats—ranging from phishing scams to viruses—that could impact them (including concerns about losing or having their phone or personal data stolen [19–21]), users do not see security strengthening of their phone as a critical concern.

In any case, the security of mobile phones is proven not to be adequate [12, 22]. Several survey studies exist that indicate in this direction. Some of these surveys studies focus on mobile phone's security issues [23] while others on mobile phone services, mentioning also security issues [16, 17, 24]. Modern smart phones, specifically, are vulnerable to more security risks [25].

As previous work has shown [2, 26], users exhibit different levels of knowledge in regard to security. Starting from the young age, they are not receiving proper cyber security and training education from schools [27], and they are lacking the security awareness and proper etiquette [28]. A method to pinpoint specific user categories and present them the right amount of information and dialogs [29] is needed in order to restore their security level.

In regard to awareness systems, there have been efforts to create a sense of accountability in a world of invisible services that we will be comfortable living in and interacting with [30] as well as mechanisms for managing security and privacy in pervasive computing environments [31] but they still focus mostly in privacy issues and not actual security enhancement through education. In any case, there are also significant legal questions as presented in [32].

In addition to the above, mobile security is not considered a critical priority by companies. Cell phone security for enterprise devices is seriously lacking, and a little misunderstood as well [33], while the majority of companies do not have a security policy that addresses mobile devices [34]. However, some initiatives are taken in the direction of protecting mobile phones against threats including policies, tools, and recruiting technically skilled personnel [35].

2.5   Conclusion

The findings of the survey that were briefly presented support the hypothesis that users can be grouped in well-defined categories according to the subjective statement of how secure they feel mobile phone communication is. These categories exhibit different values of a metric named "mean security feeling value". Further introducing a "mean actual security value", "good" security practices they follow were counted. Comparing this (objective) value to their subjective security feeling revealed interesting results. There was a clear negative connection between feeling secure and actually being secure. Users that feel that mobile phone communication is secure, tend to be less cautious in their security practices, being actually less secure than they feel. This discrepancy between user opinions on security and actual security practices is a fact that should be addressed in order to minimize vulnerabilities and user exposure.

In regards to awareness, users that feel they are very much informed believe that communication is very much secure. On the other end, users that do not feel informed are afraid that communication is not at all secure. Excessive confidence could lead to "relaxation" of security practices while excessive fear certainly hinders technology adoption and especially mobile downloading.

It is more than clear that the mobile security area is going to be the next battleground since mobile security is an emerging discipline within information security arena and security levels are not high enough [36]. Users themselves are critically affected by security and privacy threats, and play a key role in protecting themselves and others. Since they do not actively follow most of security best practices, academia and industry should focus their security awareness campaigns and efforts in order to combat the false sense of security that users have. In the following chapters of this book, we will hopefully help toward this direction, highlighting specific practical security problems.

Moreover, given the growing usage of mobile phones to access the Internet, it is of paramount importance to enhance the overall users' security levels that were found to be alarmingly low. This presents a vast opportunity for carriers and service providers too. They can play a proactive and strategic role in protecting their subscribers, both through education and through the security software they should deploy across their networks.

Manufacturers on the other hand should proceed to better designed interfaces and mobile phones generally, richer in security features. Special software could help users mitigate the security risks offering embedded encryption options for data stored in the phone as well as automated backup features and options.

Since users exhibit different levels of security feeling in regard to mobile phone communications, and since there are categories of users that face increased security risks due to their self-reassuring feeling that mobile phones are secure per se, research proposed in [37] describes a system that pinpoints and informs mobile phone users that have a low security level, thus helping them protect themselves. The system would consist of software application, installed in mobile phones as well as of software and data bases, installed in the mobile telephony operators' servers. Mobile telephony providers (by adopting this application), as well as manufacturers (by pre-installing it in their phones), could help mitigate the increased security threats effectively protecting the end users.

Software could help users mitigate the security risks associated with the usage of mobile phones. An array of tools could be implemented offering embedded encryption options for data stored in the phone as well as automated backup features and options. Of particular importance would be the implementation of software to inform the users about the encryption state of the phone, a task that seems to be a "taboo" in the mobile phone security ecosystem as we will further see in next chapter. Such software could be written either in Java, for lower-end mobile phones or in full SDK environments for the smart phones. It is also possible to collaborate with providers and implement solutions embedded in the SIM cards, using STK (SIM toolkit). The users of such a system can benefit from the following educational goals:

• Engagement and active participation in the mobile phone security field
• Understanding of the lurking dangers
• Learning how to assess the security level of the mobile phone
• Providing the tools to mitigate the dangers
• Promotion of security best practices
• Encouragement of supplying feedback
• Suggesting security actions to restore the security level

As Fig. 2.5 depicts, the system would consist of an application installed in mobile phones, and software and data bases installed in mobile operators' main servers. These applications communicate through the mobile telephony network in a

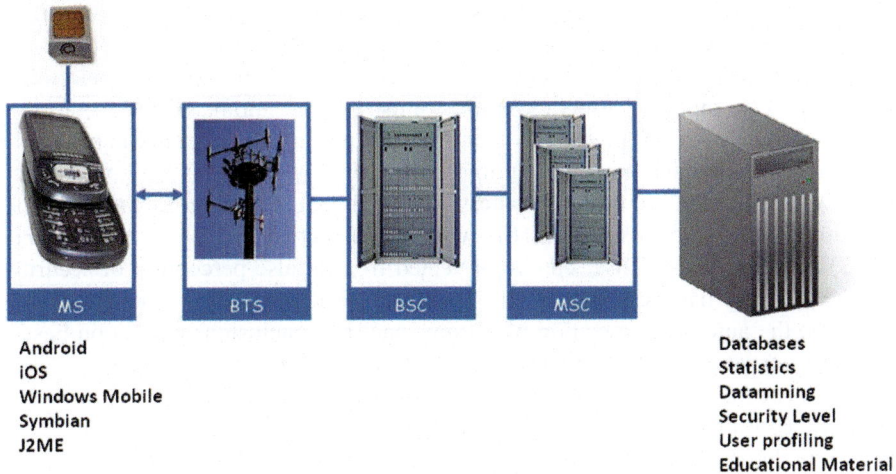

MS
Android
iOS
Windows Mobile
Symbian
J2ME

BTS

BSC

MSC

Databases
Statistics
Datamining
Security Level
User profiling
Educational Material

Fig. 2.5 System architecture

ciphered way. The mobile phone installed application (with minor differences in the array of services offered) would be able to function in all kinds of devices that have an advanced operating system (e.g. Windows Mobile, Symbian, Android, iOS). A lighter version could also be implemented for older and simple devices using J2ME (Java 2 Micro Edition).

Three main functions would be performed by the system. The first function allows pinpointing users, who have a low security level in their mobile phone, for whatever reason. The second function automatically suggests the proper methods, actions, and best practices the user has to follow in order to restore security in a higher level. Finally, the third function allows the encrypted communication and data exchange between mobile devices and provider's servers.

The device's security level evaluation function could be implemented automatically, manually, or with a combination of the two. Using the automatic method, the application transparently examines the device settings and informs the user about those that are in a state possessing security risk. In addition, by addressing questions to the user, the manual method can check aspects of his behavior that do not reflect directly to the device settings.

Furthermore, the user would be asked for his subjective opinion on how secure he feels his mobile is. As it is mentioned earlier, users can be grouped in specific security categories, based on demographical and other behavioral elements as well as on the way of using their mobile phones. Results from both the manual and the automatic method are transferred to the applications in the server, where using artificial neural networks and rules, conclusions would be extracted for the specific combination of user—mobile phone. Respectively, the answers to proper questions that examine the security practices that users follow can lead to a security behavioral prediction model of the users. It is also possible to record the hour where changes of security influencing settings take place, as to provide one more element that can help the security model.

The system would maintain data bases from studies in large user categories that provide the proper body for the system's training. These data bases would constantly be updated with the results and the metrics from the system's operations. Finally, a very important function is the comparison of automated metrics to the user's answers. As it was previously mentioned, part of user's answers can be cross-checked from the data extracted automatically. Moreover, the system could compare the subjective security awareness and feeling (according to answers for questions fifteen (15)—sixteen (16)), with the objective indicators MAAV and MASV. In this way, the user can be protected from a false perception of security that he probably has, believing that he is secure, while in reality is not. These two methods, the automatic detection of settings and the conclusions extraction based on user's answers, complete the first stage of evaluating the security level.

At the second stage, the system would implement the functionality of informing the user. Examining the current state and user's profile, the application suggests proper methods, actions, and best practices the user has to follow in order to restore (if needed) the security in a higher level.

If the device allows it and if the user accepts it, device settings could automatically be changed. Depending on the device functionality, instructions are presented to the user, either as simple text documents or as multimedia material. The user can also configure different graphical user interface and setup elements.

For the proper operation of the system, encrypted communication and date exchange between the device's application and the servers of the provider's network would take place. This communication is essential for off-loading the resource intensive neural network classification to the servers, instead of running it in the mobile device. In that way, the mobile device only records settings and the whole process takes place in the servers. Moreover, this communication allows not only the disposal of new multimedia material whenever is available but also the enrichment of the manual evaluation method with new questions when new scientific data are presented. It could also upgrade the application itself so that it can examine and locate a greater array of mobile phone's settings that reflect to its security. In any case, the communication would take place in a ciphered way so that interception is not possible.

References

1. Lenhart A (2010) Cell phones and American adults. Pew Research Center, http://www.pewinternet.org. Accessed 10 Feb 2011
2. Androulidakis I, Kandus G (2011) A survey on saving personal data in the mobile phone. In: Proceedings of sixth international conference on availability, reliability and security (ARES 2011), pp 633–638, Sept 2011
3. Androulidakis I, Kandus G. Feeling secure vs. being secure, the mobile phone user case. In: Proceedings of 7th International Conference in Global Security, Safety and Sustainability (ICGS3), Lecture Notes of the Institute for Computer Sciences 2012
4. Androulidakis I, Kandus G (2011) Correlation of mobile phone usage characteristics, security awareness and feeling to the monthly bill. In: Proceedings of the 11th International Conference on Telecommunications, June 2011, pp 257–263

5. Androulidakis I, Kandus G (2011) Mobile phone downloading among students: The status and its effect on security. In: Proceedings of 10th International Conference on Mobile Business (ICMB2011), June 2011, pp 235–242
6. Androulidakis I, Kandus G (2011) differences in users' state of awareness and practices regarding mobile phones security among EU Countries. In: Proceedings of 15th WSEAS international conference on communications, pp 296-300
7. Androulidakis I, Kandus G (2011) Ramifications of mobile phone advanced O/S on security perceptions and practices. In: Proceedings of the 3rd International Workshop on Cyberspace Safety and Security (CSS2011), pp 33–38
8. Dillman DA (1999) Mail and Internet surveys: the tailored design method, 2nd edn. Wiley, New York
9. Pfleeger SL, Kitchenham BA (2001) Principles of survey research Part 1: turning lemons into lemonade. ACM SIGSOFT Software Engineering Notes 26(6):16–18
10. Boynton PM (2004) Hands-on guide to questionnaire research: Administering, analyzing, and reporting your questionnaire. BMJ 328:1372–1375
11. Androulidakis I, Androulidakis N (2005) On a versatile and costless OMR system. WSEAS Trans Comput 2(4):160–165
12. Androulidakis I (2011) Intercepting mobile phone calls and short messages using a GSM Tester. In: Proceedings of CN2011, vol 160, CCIS. Springer, Berlin, pp 281–288
13. Synovate (2009) Global mobile phone survey shows the mobile is a 'remote control' for life, Synovate survey. http://www.synovate.com. Accessed 9 Oct 2010
14. Siau K, Shen Z (2003) Building customer trust in mobile commerce. Comm ACM 46(4):91–94
15. Leung A, Sheng Y, Cruickshank H (2007) The security challenges for mobile ubiquitous services. Inf Sec Tech Rep 12(3):162–171
16. Androulidakis I, Basios C, Androulidakis N (2007) Survey findings towards mobile services usage and M-Commerce Adoption. In: Proceedings of 18th European Regional ITS Conference, International Telecommunications Society, CD-ROM, September
17. Androulidakis I, Basios C, Androulidakis I (2008) Surveying users' opinions and trends towards mobile payment issues. Front Art Intell Appl. 169: 9–19 (Techniques and Applications for Mobile Commerce—Proceedings of TAMoCo 2008
18. McAfee (2008) Mobile security report 2008
19. Trend Micro (2009) Smartphone users oblivious to security. Trend Micro survey
20. CPP (2010) Mobile phone theft hotspots. CPP survey
21. ITwire (2010) One-third of Aussies lose mobile phones: survey. ITwire article
22. Rahman M, Imai H (2002) Security in wireless communication. Wireless Personal Comm 22(2):218–228 [Online]
23. Androulidakis I, Papapetros D (2008) Survey Findings towards awareness of mobile phones' security issues, recent advances in data networks, communications, computers. In: Proceedings of 7th WSEas international conference on data networks, communications, computers (DNCOCO '08), Nov. 2008, pp 130–135
24. Vrechopoulos AP, Constantiou ID, Sideris I (2002) Strategic marketing planning for mobile commerce diffusion and consumer adoption. In: Proceedings of MBusiness 2002, July 8–9
25. comScore M:Metrics (2008) Smarter phones bring security risks: Study. http://www.comscore.com. Accessed 9 Oct 2010
26. Allam SA (2009) Model to measure the maturity of smart-phone security at software consultancies, Thesis, University of Fort Hare. http://hdl.handle.net/10353/281
27. National Cyber Security Alliance (NCSA) (2009) Schools lacking cyber security and safety education
28. Cable & Wireless (2009) Workers lack mobile phone etiquette
29. De Keukelaere F, Yoshihama S, Trent S, Yu Z, Luo L, Zurko ME (2009) Adaptive security dialogs for improved security behavior of users, human-computer interaction—INTERACT 2009, LNCS 2009, Vol 5726. Springer, Heidelberg, pp 510-523
30. Langheinrich M (2002) A privacy awareness system for ubiquitous computing environments. In: Proceedings of UbiComp, pp 237–245

31. Cornwell J, Fette I, Hsieh G, Prabaker M, Rao J, Tang K, Vaniea K, Bauer L, Cranor L, Hong J, McLaren B, Reiter M, Sadeh N (2007) User-controllable security and privacy for pervasive computing. In: Eighth IEEE workshop on mobile computing systems and applications, HotMobile 2007

32. Nancy J. King NJ, Jessen PW (2010) Profiling the mobile customer—Privacy concerns when behavioral advertisers target mobile phones. Computer Law Security Rev 26(5):455–478

33. ABI Research (2009) Study: enterprises need to address cell phone security

34. TechRepublic (2007) Survey respondents say companies are lax on mobile security. TechRepublic article

35. Darkreading (2010) Survey: 54 percent of organizations plan to add smartphone antivirus this year. Darkreading article

36. Goode Intelligence (2009) Mobile security the next battleground. http://www.goodeintelligence

37. Androulidakis I, Kandus G (2012) PINEPULSE: A System to PINpoint and Educate Mobile Phone Users with Low Security. In: Proceedings of 7th International conference in global security, safety and sustainability (ICGS3), Lecture notes of the institute for computer sciences, vol 99, pp 62–66

Chapter 3
Voice, SMS, and Identification Data Interception in GSM

Abstract In this chapter, the reader will get an insight into one of the most easily employed techniques of voice, SMS, and identification data interception in GSM networks. Using a fake base station that mimics the behavior of a legitimate base station of the mobile phone operator, a malicious entity can convince mobile phones in a given area to handle their communication to it, effectively launching a man in the middle attack. This attack is possible only in GSM networks, since 3G employs mutual authentication, where the base station too has to authenticate its validity to the handset. However, it is relatively easy to use a jammer, jamming the 3G band. Almost every single mobile phone nowadays is multiband capable and as such it will fall back to GSM operation where it can be intercepted using the fake base station method.

Keywords GSM eavesdropping • Voice interception • IMSI catcher • SMS interception • Man in the middle • UMTS jammer • GSM tester • GSM repeater • Ciphering Indicator • OFM bit

3.1 Introduction

Interception of mobile phone calls used to be a simple radioscanning exercise back in the era of the first analog systems. Digital systems such as GSM proved to be a lot more secure, encompassing encryption and difficult to overcome complexity. As it is the case with every other technology, the scientific community soon started theoretical discussions about the algorithms' security. Following, attackers managed to mount practical attacks too. There are many papers discussing crypto attacks to the GSM standard itself or to its various implementations by different vendors (as briefly discussed in the first chapter) but in this book we will focus into practical issues and not theoretical or algorithmic discussions.

GSM mobile phone communications can easily be intercepted, without performing any cryptanalysis, using a fake base station. One of the fundamental security problems and a basic shortcoming of the GSM security planning was the fact that the mobile telephony network does not have to authenticate itself to the user. Only the user has to authenticate himself in order to gain access to the network. It is well known that a user wishing to gain access in a provider's GSM/UMTS mobile telephony network must own the proper SIM card (protected with the PIN code) inserted in his phone device. The user's legitimacy is therefore checked by comparing the SIM's credentials with the data saved in the network's database. This way, the user is authenticated in the network and can use its services. In GSM networks, this basic principle of authentication is not implemented in respect to the side of the provider. Base stations do not employ any identity authentication mechanism. Respectively, mobile phones are not capable to assess and certify the legitimacy of the system they are connecting to and whether this system is indeed part of their provider's network. On the other hand, UMTS employs mutual authentication, where the base station too has to authenticate its validity to the handset. However, it is relatively easy to use a jammer, jamming the 3G band. Almost every single mobile phone nowadays is multiband capable and as such it will fall back to GSM operation where it can be intercepted using the fake base station method.

Secondly, encryption is not mandatory and, if present, the specific algorithm to be used can be negotiated. In case the base station does not support any encryption algorithms, following negotiation with the mobile phone, the call can proceed without encryption. This way, a fake base station in the proximity of a user, without encryption supported (or specifically disabled), is all that is needed to intercept the communication using a simple man in the middle approach. In man in the middle attacks, as the name suggests, the malicious entity is placed between the original callers. If caller A was going to call B and C is the malicious entity what happens is this: Instead of the direct flow of information from A to B, there is a flow of A to C and then from C to B. Caller A thinks she is talking to B while she is actually talking first to C. Malicious entity C of course relays the communication back to B so the attack succeeds.

Thus, the only thing an attacker has to do is activate a fake base station in a given area, pretending that it is part of the network of the victim's provider. One of the basic characteristics of GSM proves to be a strong ally in this effort: Each mobile phone constantly monitors a special data transmission channel—beacon (BCCH—broadcast control channel) from the nearby base stations in order to choose for its communication the one offering the best characteristics (usually the closest one). This way the device achieves great economy in the consumed energy by transmitting in lower power and increases its autonomy time and quality of speech. Hence, should the attacker install in a given area his equipment and start transmitting, overlapping in power the authentic base stations' signals, mobile phones located close by will rush to connect with him.

The next stage in the attack is the encryption's neutralization. GSM uses algorithm A5 for voice encryption. There exist various versions of this algorithm that offer different levels of security (A5/2, A5/1, A5/3—sorted in strength order from lowest to highest). There is also a no encryption at all version (A5/0). Under normal circumstances, the network has stored in its authentication center's database in the Home Location Register (HLR) the secret key Ki which is also stored in the user's SIM card and is never transmitted in the network. Since Ki is never transmitted, the network challenges the SIM sending a random 128 bit number (RAND). Using algorithm A3, the SIM produces a 32 bit Signed Response (SRES) that is transmitted back to the network. It is this value that is being compared to the respective one in the HLR. Should these values match, the SIM is authenticated. Finally, using Ki

Table 3.1 Authentication and encryption algorithms

A3	Takes the 128 bit Subscriber Authentication Key (Ki) that is stored into SIM and to the HLR and produces a 32 bit Signed Response (SRES) answering to a random 128 bit number (RAND) which is sent by the HLR.
A8	Produces a 64 bit Session Key (Kc) from the 128 bit random number (RAND) and the 128 bit Ki.
A5	Uses Kc and the sequence number of the transmitted frame and cryptographs the speech. A5 is implemented into the phone.

and RAND, algorithm A8 produces the session key Kc which is fed to the speech encryption algorithm A5. Table 3.1 presents these algorithms. Quite interestingly, both A3 and A8 algorithms were implemented in a single algorithm, COMP128.

In the case of the fake base station, the Ki key is not known to the attacker. However since he is the "network" he can accept whatever SRES the mobile phone sends. The SIM (and the mobile phone) will believe that it was properly authenticated. The attacker still needs Ki to derive Kc in order to decrypt the voice transmission that will follow. Once again, system planning prioritizes usability instead of security. The corresponding protocols allow the negotiation and the agreement between the mobile phone and the base station regarding whether they will use an encryption algorithm, and if so, which one. Using the proper signaling, the fake base station informs the mobile phone that it does not have any encryption capability (A5/0) and thus the mobile phone will start communicating with it without using any encryption at all. From that point on, using the proper equipment, it is a trivial task to demodulate and record the digital communication that will follow.

Another characteristic that facilitates these attacks is the fact that they are usually targeted, since a particular mobile phone target is being intercepted. Therefore a few milliwatts of power are enough for a base station located 20–30 m away from the mobile phone to overpower the legitimate provider's transmission power that is located some hundred meters away. Since the fake base station seems to be a better choice for the mobile phone, it will happily hand over its communication to it.

Up to this point, we have described the logic that the fake base station uses to "capture" the communications of mobile phones that are located in his coverage area. In order to realize the call, a connection with the normal network is needed. So acting as a man-in-the-middle, the attacker interconnects his system with the rest of the network, using a simple mobile or normal land line telephone that relays the communication back to the genuine network and his intended initial recipient. Needless to say, a thorough recording of the conversation will take place in the interceptor's systems apart from relaying it.

3.2   Practical Setup and Tools

The necessary setup to perform this attack ideally consists of an industrial base station, the same as the ones used by the network operators. However, this expensive setup can easily be substituted using advanced GSM testing equipment.

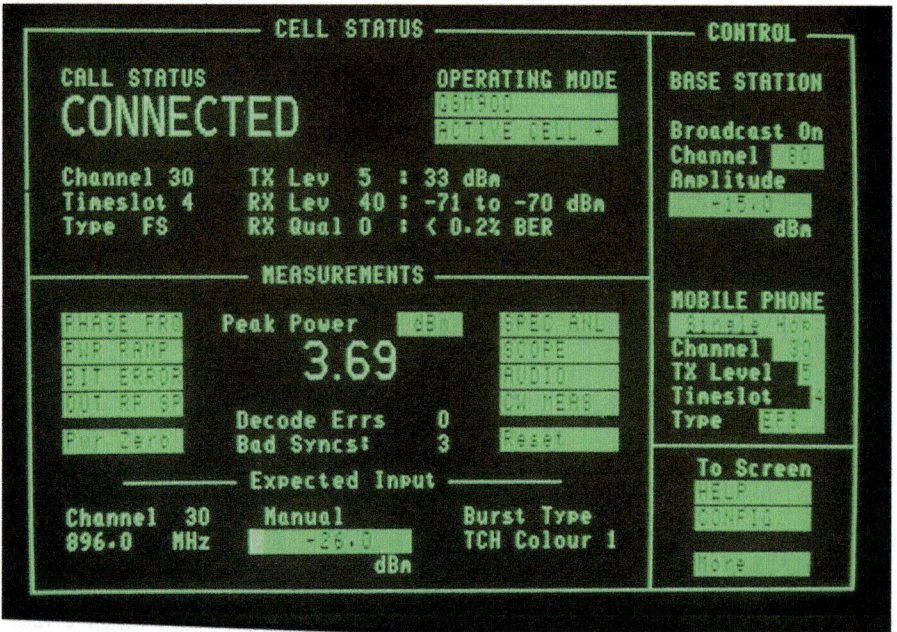

Fig 3.1 Typical measurements performed by the tester

Such equipments provide all the necessary signaling for the operation of handsets and can also demodulate voice from the digital signal transmitted in the air.

During their connection with the mobile phone they are performing a wealth of measurements, pinpointing even the slightest problem and malfunction in the operation of the phone (Fig. 3.1). They can also intercept the IMSI (International Mobile Subscriber Identity) of the user's SIM (Subscriber Identity Module) card and the IMEI (International Mobile Equipment Identity) of his handset. The ability to intercept IMSI has led to the name "IMSI catcher" being collectively used to describe this family of equipment. At the same time, they can read the short messages (SMS) that the user is trying to send. The malicious user can further initiate calls or send messages to the victim, choosing freely any caller identity he pleases.

A very interesting fact is that not only plaintext messages but also specific binary ones can be send toward the handset. Such messages are normally restricted only for provider's usage and blocked in case a simple user tries to send them. Using this equipment, this check is circumvented and the attacker can indeed send such messages.

It must be noted that this attack works only for outgoing calls. The original network can't locate the handset anymore since it is camped in a base station not belonging to the operator and as such can't terminate calls to it. Furthermore, as stated before, although UMTS networks employ mutual authentication, they can be jammed, forcing the mobile phone to use GSM and subject to this interception technique.

In the past few years, research in open source software and hardware has led to the realization of "private" base stations, running open source software. Costing a fraction of the cost of a real base station, they implement most of its functionality, proving a valuable tool for research (and possibly fraudsters too). So, apart from a GSM tester, it is also possible to use hardware and software from the open-source projects OpenBTS. It is a project that uses a software radio to present a GSM air interface to standard GSM handset and uses a SIP softswitch or PBX to connect calls. There is also OpenBSC that implements higher level functionality and needs a hardware BTS to operate. For the purposes of this book, we will continue with the description of the attack using a GSM tester only, which is considerably easier to operate. In contrast to GSM testers, which can be operated with minimal training, the use of these open source tools is relatively more difficult. Interested and more advanced level readers can find more details about OpenBTS and OpenBSC in [1, 2].

In order for the attacker to launch the attack, the following tools are needed:

(a) A GSM tester [3–5] such as the one seen in Fig. 3.2.
(b) An antenna (GSM testers are usually connected with a special cable directly to the handset under test, but in our case the attacker transmits using the antenna).
(c) A GSM repeater [6] (optionally, in order to increase the effective distance of the interception).
(d) A mobile handset with monitoring software installed or enabled (further explained below).

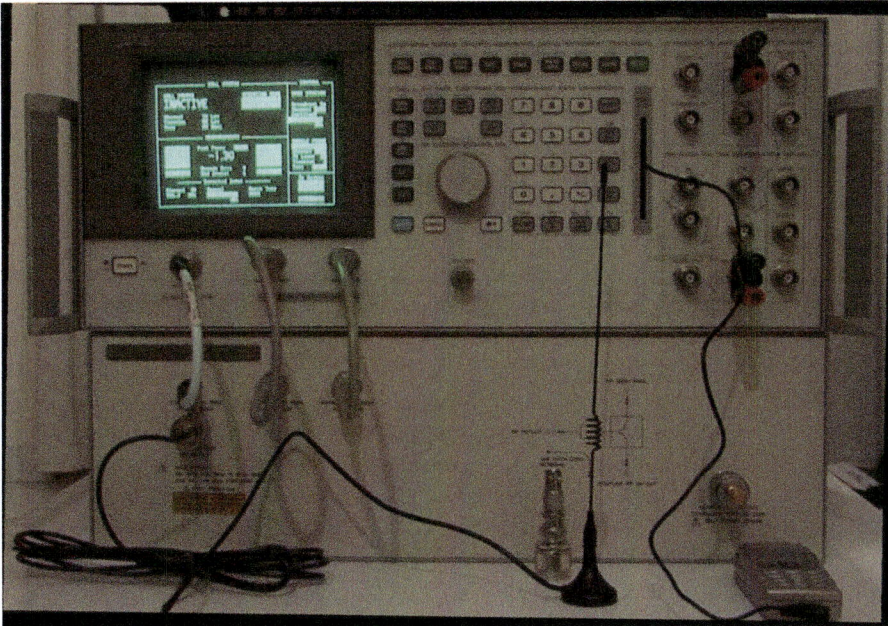

Fig. 3.2 A typical GSM tester

(e) A second mobile phone or land line in order to "channel" the interception communication through it. This phone is connected to the audio in/out outputs of the demodulator, as seen to the right of Fig. 3.2.

A PC with serial port or other connectivity would enable the full automation of the process. In any case, the connection details are straight forward and need not be presented here.

3.3 Implementation

Following the proper connection of the equipment, parameterization has to take place. The required data that should be set are first of all the country number (MCC—Mobile Country Code) and the mobile telephony network number (MNC—Mobile Network Code) of the SIM of the mobile phone to be intercepted. ITU-T E.212 [7] defines a list of Mobile Country Codes (MCCs) for use in identifying the country of origin of mobile stations in wireless telephone networks, particularly GSM and UMTS networks. There are also Mobile Network Codes (MNCs) that further discriminate operators in a given country. As such the MCC/MNC combination is universally unique for each operator. This information is freely available and can even be accessed by Netmonitor or other diagnostic tools. Table 3.2 shows the relevant entries for Greece.

Proceeding, Netmonitor (or Engineering Menu or Field Test Display) will be used to gather the original network's operating parameters. In the next chapter, we will further describe it, but for the moment it is enough for the reader to know that it is a special mode in mobile phones, used to measure network and phone operating parameters and status [8]. When activated, a new, additional menu usually appears, providing a wealth of information.

The most important clue that Netmonitor provides for the needs of the attack is the ARFCNs (absolute RF channel numbers) that nearby GSM base stations BCCHs (broadcast control channels) transmit in. In Fig. 3.3, such a Netmonitor phone screen is presented, providing information about BCCHs in the vicinity of the phone. BCCH is a signaling channel that carries information about the identity, configuration, and available features of the base station as stated in the introduction of the chapter. Mobile phones continuously "listen" to that broadcast signal in order to be able to communicate with the GSM network. This channel also provides a list of ARFCNs used by neighboring base stations. Since the mobile phone is monitoring a specific list of channels each time, the attacker must chose one of these channels in order to transmit, deceiving the handset. This way he will effectively be masking the legitimate signals and seizing control of the nearby mobile phones.

Table 3.2 MCCs and MNCs for Greece	MCC	Country	MNC	Operator
	202	Greece	01	Cosmote
	202	Greece	05	Vodafone
	202	Greece	10	WIND

Fig. 3.3 Netmonitor showing the neighboring channels

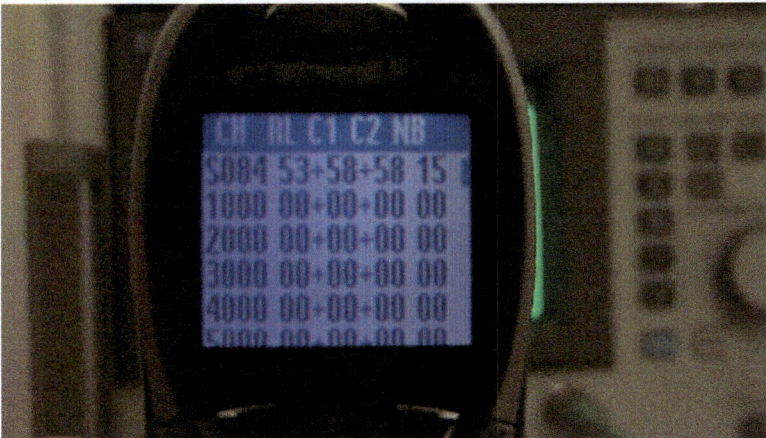

Fig. 3.4 Victim phone has camped on to the fake base station

As it can be seen in Fig. 3.4, a few moments after the operation of the GSM tester, the "victim" phone camps on to the fake base station according to GSM standards procedures [9–11]. Namely, the GSM tester is transmitting in BCCH channel 84 (which was the already selected channel by the mobile phone before the "intrusion" as was depicted in Fig. 3.3). The received signal from the GSM tester overpowers the legitimate signal since the transmitter is far closer than the antenna of the base station of the network provider. It is also interesting to note that in the specific setup the BCCH is instructed not to advertise any other BCCHs (hence the rest of neighboring channels is full of 00). As such, even if the phone moves closer to another base station, it will not switch to that, unless the signal of the fake base station is completely lost. This way, the attacker can easily keep the connection with the victim which otherwise would be lost the moment the mobile phone

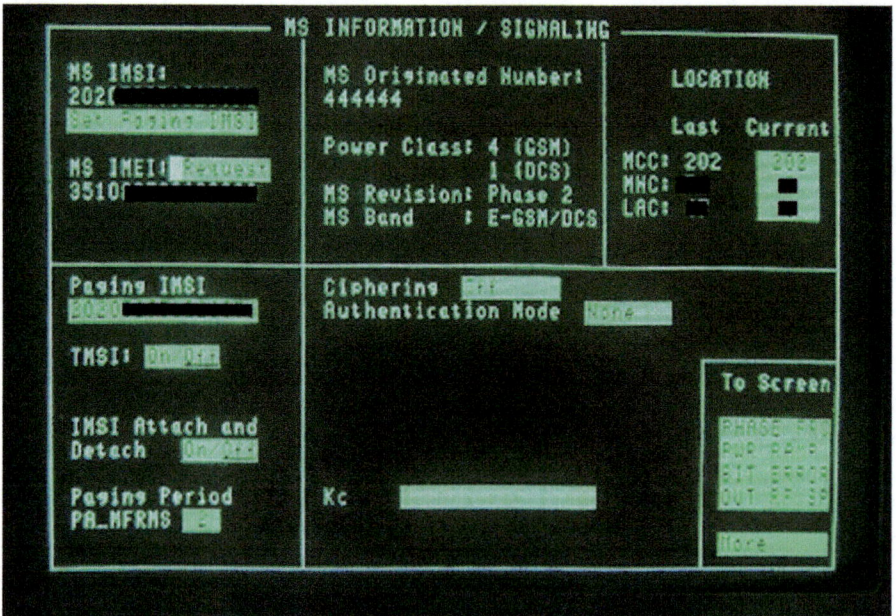

Fig. 3.5 Details of IMSI, IMEI, and calling number of the victim, available to the GSM tester

entered an area served by a base station with better characteristics than the fake one. The final step is to deactivate the encryption by choosing A5/0 algorithm as described earlier.

After this point, every call attempt originating from the cell phone will be logged by the GSM tester. In Fig. 3.5, we can see, among other things, that IMSI, IMEI, and the number that the user is trying to dial are decoded. One has merely to dial the number requested (444444 in our example) by the "victim" phone, using the second phone which will actually dial the call, channeling the communication, as in the classical concept of the man in the middle attack.

3.4   Problem Solution

3.4.1   In General

Having described the attack, in the coming paragraphs, we will see ways of protection against fake base stations, the behavior of mobile phone's Graphical User Interfaces in regard to this attack, and a small revision of the history of the respective standards.

3.4.2 History of the Ciphering Indicator

GSM engineers suggested that the lack of encryption, being a telltale that the phone is under a possible interception, should be mentioned to the user. This could be achieved with a special indicator. As we will see, it apparently took many years for that mandatory feature to mature and unfortunately, even today, not all manufacturers comply with it. Most of the handset vendors implement this mechanism using an icon or a cryptic symbol that the user has to figure out its meaning (e.g. an exclamation mark or an unlocked pad). As expected, users themselves (more than 75 % of them) are completely unaware of the existence of this indicator [12]. Making things even worse, the standard allows the operator to disable this feature in the SIM. So, even if a mobile phone adheres to the standard and has a special indicator, the operator can opt to disable it. As such the user will never be informed when encryption is switched off.

During the long history of GSM standards, if one were to have a look at prETS 300 977 (GSM 11.11 V5.5.0) [13] in May 1997, he would have noticed in Elementary File 6FAD (the equivalent of data files in SIM as we will see in Chap. 6), Paragraph 10.3.18, the notion of a "cryptic" OFM bit (Fig. 3.6). The meaning of OFM wasn't mentioned in the abbreviations nor elsewhere in the text.

Identifier: '6FAD'	Structure: transparent		Mandatory
File size: 3+X bytes		Update activity: low	

Access Conditions:
    READ   ALW
    UPDATE   ADW
    INVALIDATE   ADW
    REHABILITATE ADW			

Bytes	Description	M/O	Length
1	MS operation mode	M	1 byte
2 - 3	Additional information ⬅	M	2 bytes
4 - 3+X	RFU	O	X bytes

Byte 3:

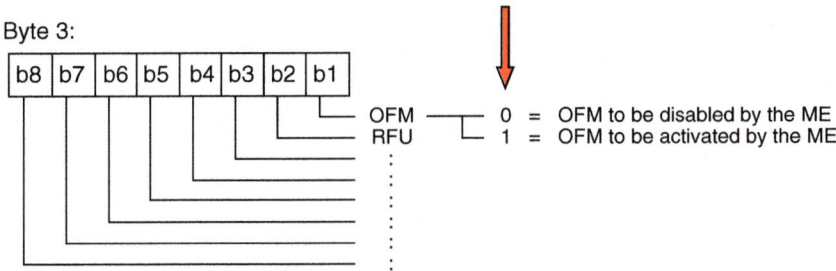

Fig. 3.6 The first occurrence of OFM bit in the standards

A few months later, in July 1997, GSM 02.09 V4.4.0 [14], Paragraph 3.3.3 states for first time the "ciphering indicator," referencing back to GSM 11.11, which by the way, still used the term OFM at the moment (ME is the mobile equipment, practically the mobile phone):

> "The ME has to check if the user data confidentiality is switched on using one of the seven algorithms as defined in GSM 02.07. In the event that the ME detects that this is not the case, or ceases to be the case (e.g. during handover), **then an indication is given to the user. This ciphering indicator feature may be disabled by the SIM** (see GSM 11.11).
> In case the SIM does not support the feature that disables the ciphering indicator, then the ciphering indicator feature in the ME shall be enabled by default.
> The nature of the indicator and the trigger points for its activation are for the ME manufacturer to decide. During the establishment of a call the trigger point shall be at call initiation at the latest. In the case of handover the trigger point shall be the completion of handover at the latest.
> The manufacturer may provide the means to enable the user to temporarily disable the feature. This should be done in such a way that the user can protect it from misuse"

Then, in January 1998, there is an addition in ETS 300 505 (GSM 02.07 version 4.8.2) [15], Paragraph B.1.22 that restates the functionality of the ciphering indicator:

> "The ciphering indicator feature allows the ME to detect that ciphering is not switched on and to indicate this to the user, as defined in GSM 02.09.
> The ciphering indicator feature may be disabled by the home network operator setting data in the "administrative data" field (EFAD) in the SIM, as defined in GSM 11.11.
> If this feature is not disabled by the SIM, then whenever a connection is in place, which is, or becomes unenciphered, an indication shall be given to the user.
> Ciphering itself is unaffected by this feature, and the user can choose how to proceed;

Almost 2 full years later, for the first time, the OFM bit gets "explained" in the original 11.11 standard too. In version GSM 11.11 V8.1.0 [16] (November 1999), the abbreviations list gets amended to include the definition of the term as "Operational Feature Monitor". Furthermore, as one reads in Paragraph 10.3.18: "The OFM bit is used to control the Ciphering Indicator as specified in GSM 02.07". This was also the first time where the OFM bit was connected to its initial meaning as described in GSM 02.07.

Descendants of GSM 11.11 are 3GPP 51.011 and then 3GPP 31.102. In 3GPP 31.102 V6.5.0 [17] (March 2004), the notion of OFM gets at last abandoned in the description of EFAD in Paragraph 4.2.18 in favor of the more straight-forward term of Ciphering Indicator Feature (Fig. 3.7). It is also interesting to note that reference no 17 of that standard still points to 02.07 although it is not mentioned anywhere in the text anymore.

Byte 3(second byte of additional information):

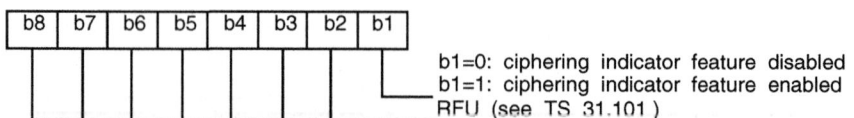

b8	b7	b6	b5	b4	b3	b2	b1

b1=0: ciphering indicator feature disabled
b1=1: ciphering indicator feature enabled
RFU (see TS 31.101)

Fig. 3.7 OFM gets abandoned in 3GPP 31.102 V.6.5.0

Regarding GSM 02.07, its descendant is 3GPP 22.101 where one first finds the first occurrence of Ciphering Indicator in V3.8.0 (December 1999) [18]. Finally, in 3GPP 42.009 (descendant of GSM 02.09), there are no changes spotted in the relevant section describing ciphering indicator (Sect. 3.3, User data confidentiality on physical connections (Voice and Non-voice)—Paragraph 3.3.3, Functional requirements).

3.4.3   Recent Additions

As described, even if the mobile has implemented the feature, the operator can disable the ciphering indicator. 10 years after 3GPP's 22.101 first occurrence of Ciphering Indicator, in March 2009, with 3GPP 22.101 V8.11.0 [19] there are some "Clarification and enhancement of ciphering indicator feature" as stated in the new phrasing, in Paragraph 14:

> Types of features of UEs The ciphering indicator feature allows the UE to detect that the 3GPP radio interface ciphering (user plane) is not switched on and to indicate this to the user. The ciphering indicator feature may be disabled by the home network operator setting data in the SIM/USIM. The default terminal behaviour shall be to take into account the operator setting data in the SIM/USIM. However, terminals with a user interface that can allow it, shall offer the possibility for the user to configure the terminal to ignore the operator setting data in the SIM/USIM. If this feature is not disabled by the SIM/USIM or if the terminal has been configured to ignore the operator setting data in the SIM/USIM, then whenever a user plane connection is in place, which is, or becomes un-enciphered, an indication shall be given to the user. In addition, if this feature is not disabled by the SIM/USIM or if the terminal has been configured to ignore the operator setting data in the SIM/USIM, then additional information may be provided about the status of the ciphering. Ciphering itself is unaffected by this feature, and the user can choose how to proceed;"

3.4.4   The Actual Situation

This last standard seems to be the first step toward actually empowering the user to overcome the control of the operator, in regard to the ciphering indicator. Sadly, recent research [20] has not found any handsets that implement the feature of manually configuring the terminal to ignore the operator's setting of ciphering indicator.

During this research, a test network with encryption switched off was used. A test network with MCC=001 and MNC=01 is provisioned in the standards as to not interact with other networks during tests. Different manufacturers' approaches were found, testing a big array of mobile phones, ranging from old to new ones and from simple to smart phones. There were cases where the ciphering indicator was completely absent, presented as an icon, as a message only and as a message plus an icon.

The indicators/icons themselves (marked with an arrow in the relevant Figures) manifested as stars and an exclamation mark (Fig. 3.8), as an unlocked padlock

Fig. 3.8 Stars with the exclamation mark in the middle indicate (to a savvy user) the lack of encryption

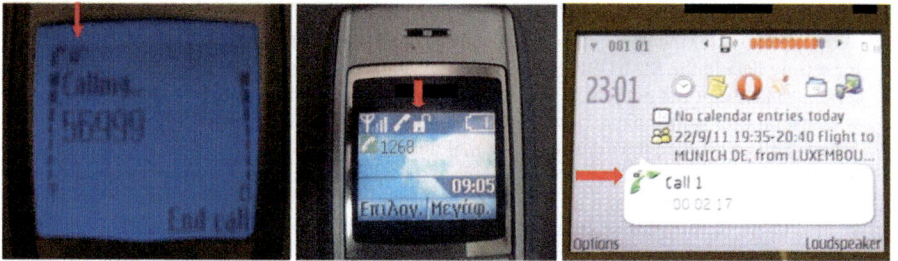

Fig. 3.9 The unlocked padlock indicates (a little more descriptively) the lack of encryption

(Fig. 3.9) or as a red triangle or gray square with an exclamation mark inside. Certain mobile phone manufacturers advanced a step further flashing a proper warning message to the user (Fig. 3.10).

This informatory message in most cases would disappear after a couple of seconds, while in the more robust implementations it would stay until the user specifically acknowledged reading it. Other manufacturers, unfortunately, do not implement the ciphering indicator mechanism at all (although mandatory by the standard). It also seems that modern smart phones do not employ this mechanism. In our tests, most mobile phones with the leading modern operating systems failed to show any kind of indication when encryption was switched off. The owner of such a mobile phone will not be informed about the lack of encryption even if the corresponding bit in the SIM card is set. Quite interestingly, there were cases in older phones where the icon popped up even if the relevant ciphering indicator bit was set to 0 in the SIM file, possibly due to a bug. In another brand, older phones

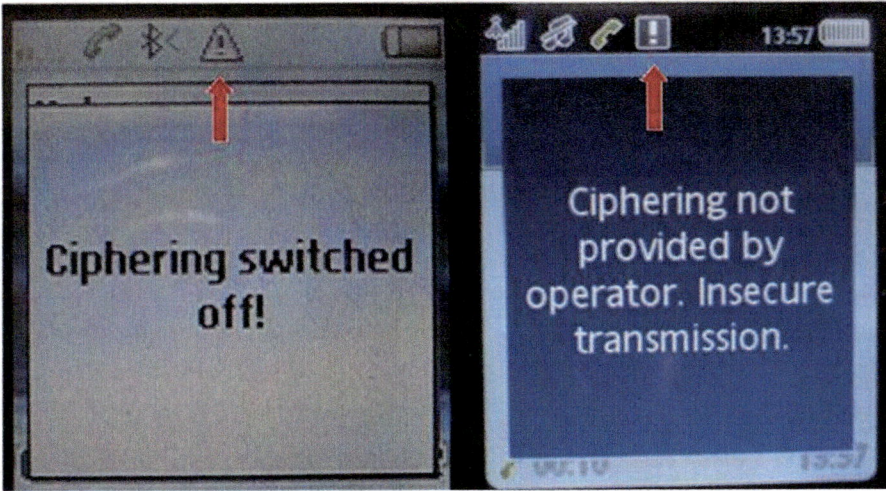

Fig. 3.10 A far better implementation, the triangle with the exclamation mark, remains lit during the call while a clear message stating the lack of encryption is flashed for a few seconds after the beginning of the call

with small displays actually used a bigger icon than currently used in the colorful large displays. The current icon is so small that it is difficult for the user to notice.

Another significant issue is whether the Ciphering Indicator icon is documented in the phones' manuals. The way the icon of the manufacturer using the approach of flashing a message is explained in the phone's manual as shown in Fig. 3.11. However, proper documentation for the Ciphering Indicator was rather rare, and only one manufacturer in the considered data set included it. Another manufacturer has documented, in three different cases, that a closed padlock icon shows that data services (e.g., WAP and WiFi) use encryption and that the absence of this icon indicates a lack of encryption, but they have provided no documentation at all regarding the open padlock icon for voice and text communication.

At this point, it is interesting to examine the behavior of the so-called smartphones. Taking into consideration the three main operating systems (Android, iOS, and Windows Mobile) as well as the older Symbian, it was found that only Symbian had implemented a Ciphering Indicator. This could possibly be attributed to the fact that Symbian had originally close ties with a manufacturer that has been using a Ciphering Indicator since its early models, whereas the other advanced operating systems have evolved out of the computer community (i.e., Windows Mobile, Android, and iOS). It should also be noted that in the case of Android, a similar icon with the implementation of the "Ciphering Indicator" (a triangle with an exclamation mark) is used as a general notification icon, as shown in Fig. 3.12, but it is not used to alert the user when the encryption is switched off. It is an altogether different icon.

These results show that graphical user interfaces have not evolved at all in regard to security and that many manufacturers still do not provide the necessary

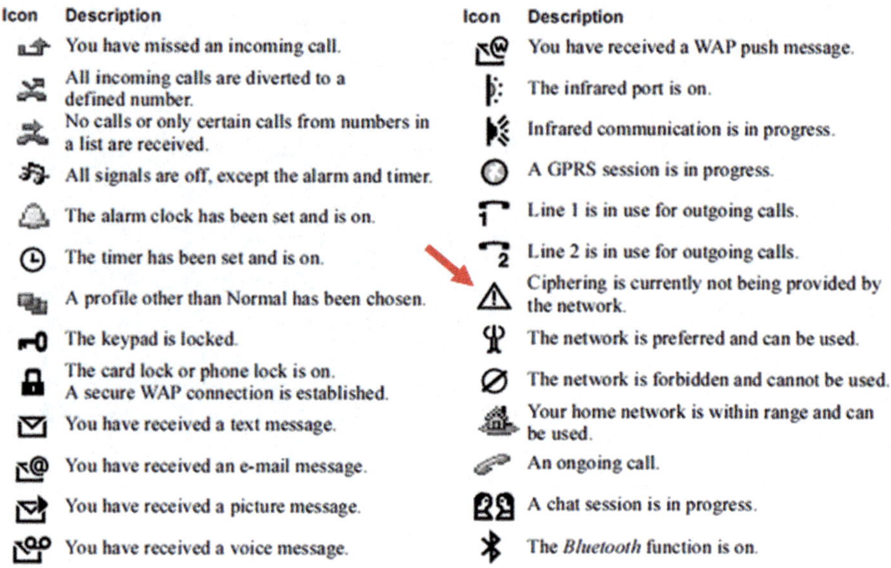

Icon	Description	Icon	Description
	You have missed an incoming call.		You have received a WAP push message.
	All incoming calls are diverted to a defined number.		The infrared port is on.
	No calls or only certain calls from numbers in a list are received.		Infrared communication is in progress.
	All signals are off, except the alarm and timer.		A GPRS session is in progress.
	The alarm clock has been set and is on.		Line 1 is in use for outgoing calls.
	The timer has been set and is on.		Line 2 is in use for outgoing calls.
	A profile other than Normal has been chosen.		Ciphering is currently not being provided by the network.
	The keypad is locked.		The network is preferred and can be used.
	The card lock or phone lock is on. A secure WAP connection is established.		The network is forbidden and cannot be used.
	You have received a text message.		Your home network is within range and can be used.
	You have received an e-mail message.		An ongoing call.
	You have received a picture message.		A chat session is in progress.
	You have received a voice message.		The *Bluetooth* function is on.

Fig. 3.11 An example of the documented Ciphering Indicator

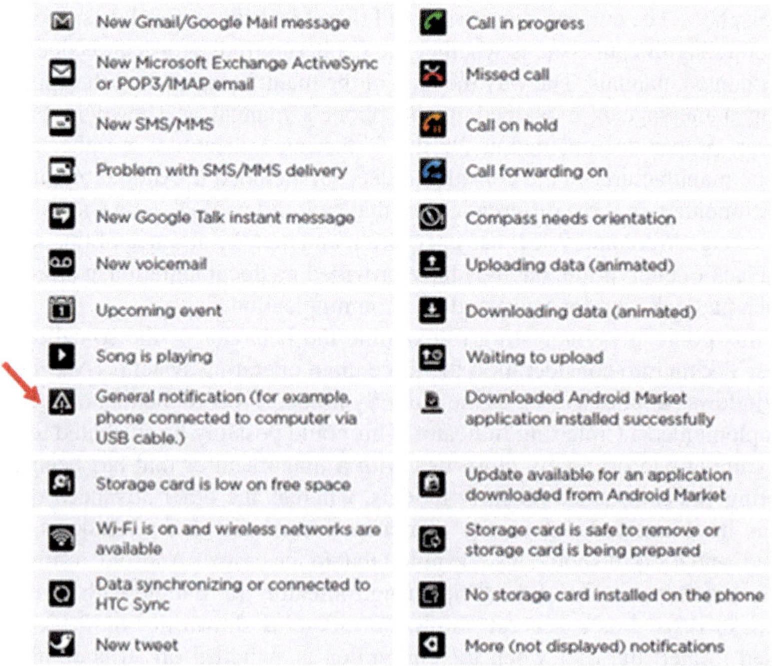

Icon	Description	Icon	Description
	New Gmail/Google Mail message		Call in progress
	New Microsoft Exchange ActiveSync or POP3/IMAP email		Missed call
	New SMS/MMS		Call on hold
	Problem with SMS/MMS delivery		Call forwarding on
	New Google Talk instant message		Compass needs orientation
	New voicemail		Uploading data (animated)
	Upcoming event		Downloading data (animated)
	Song is playing		Waiting to upload
	General notification (for example, phone connected to computer via USB cable.)		Downloaded Android Market application installed successfully
	Storage card is low on free space		Update available for an application downloaded from Android Market
	Wi-Fi is on and wireless networks are available		Storage card is safe to remove or storage card is being prepared
	Data synchronizing or connected to HTC Sync		No storage card installed on the phone
	New tweet		More (not displayed) notifications

Fig. 3.12 General notification icon in Android: triangle with exclamation mark

means to adequately inform users. In the best of cases, that information is just an icon, usually not documented in the manual. Other popular brands and most of the modern smart phones do not even employ a mechanism to alert the user about the lack of encryption.

3.5 Software-Defined Radio and Modified Firmware

As noted in Chap. 1, up to recently, it has been traditionally difficult to intercept air waves for mobile phone signals and to interact with mobile phones transmitting on their radio layer. Some open source projects during the past few years, such as GnuRadio [21], AirProbe [22], OpenBTS [23], OsmoBTS [24], OpenBSC [25], and OsmocomBB [26] along with commercial or do-it-yourself open source software-defined radio peripherals such as universal software radio peripheral series (USRP) [27], HackRF [28], bladeRF [29] , sysmoBTS [30] , UmTRX [31], have brought this functionality to the interested user with great success. In the same wavelength, Myriad-RF is a family of open source hardware and software projects for wireless communications, and a community that is working to make wireless innovation accessible to as many people as possible.

In software-defined radio (SDR) instead of using certain hardware blocks to achieve a radio communication system, software means are used, offering an unparalleled versatility and ease of development. There exist various SDR-based hardware boards that can be used in order to transmit and receive GSM signals. Such hardware is USRP, the development kit from Range Networks, UmTRX, and sysmoBTS. Their cost lies in the range of a few thousands of euros, which makes them accessible to a large base of interested individuals.

Especially for creating a GSM network (that can probably be used for fraudulent activities) using open source SDR there are two main software options: OpenBTS and OsmoBTS. OpenBTS is a Unix application that uses a software radio to present a GSM air interface to standard 2G GSM handset (such as a BTS does) and uses a SIP softswitch or PBX to connect calls. OsmoBTS is a software implementation of Layer2/3 of a BTS.

In addition, the OpenBSC and Osmo-BSC [32] provide software that can be used as a traditional GSM switching centre (BSC), while if used in conjunction with BTS hardware can be and configured as a fully functional GSM network. OpenBSC further implements a minimal subset of the BSC, MSC, and HLR so that it can be used as a "network in the box," while Osmo-BSC runs in pure BSC-only mode, i.e., it will implement the BSC functionality and the A-bis and A interfaces, but not any further protocols (Fig. 3.13).

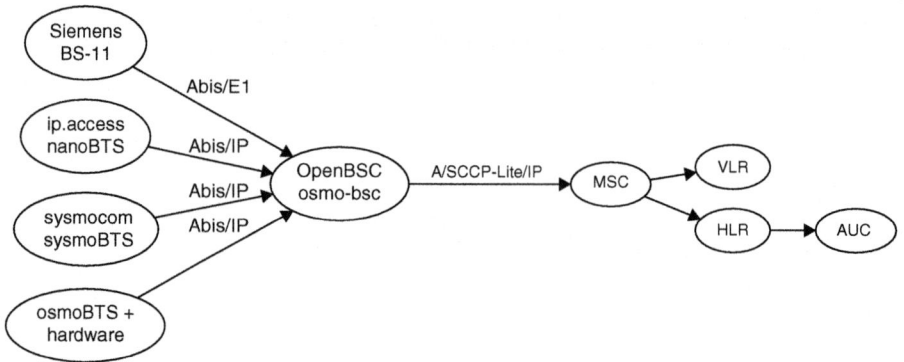

Fig. 3.13 taken form osmocom.org depicts the interaction of these elements toward building a functional GSM network

3.6   Conclusion

As it was presented in this chapter, a malicious user with the proper equipment and minimal knowledge of the network details can relatively easily intercept conversations and short written messages of nearby mobile phones. He can also place calls to his victims or send them messages choosing any phone number identification he pleases, effectively masquerading himself. This specific attack exploits the lack of mutual authentication and it is difficult to be amended in the present GSM's implementation. Network authentication is applied in 3G networks, and thus the user is protected effectively from this specific attack (although backward compatibility with GSM can circumvent the security measures). Taking into consideration these facts, there are at least two ways to mitigate the problem. The first one involves clear and easy to understand messages and icons, informing the user about the situation. The second, in regard to the protocol suite, is to have a mandatory authentication of the network to the handset as is the case with 3G networks.

Unfortunately, only a few devices have some sort of notification mechanism/ indication for the security level. Despite that, notification is under the provider's control via the SIM card's parameterization (which can also take place remotely without users realizing it). Moreover, in certain mobile phone devices, this indication is presented with incomprehensible symbols.

Besides the built-in mechanism, the user can always activate/install the Netmonitor application in his mobile phone and receive feedback for the provided encryption level (as we will shortly see in next chapter). This solution, albeit not very practical, is foolproof because it does not fall into network's control and can reveal the level of encryption used regardless of the operator's settings.

Another clue that something is wrong is the fact that, while the victim's phone is under surveillance it cannot accept any calls or messages since it is effectively cut out of its home network. The quality of the voice might also give a hint since it will usually be worse than normal.

Since mobile phones are becoming targets of modern cyber criminals, it is necessary to educate the users and to raise the public's awareness. Mobile phone manufacturers on the other hand have to actively participate in this effort, developing user interfaces that will clearly inform the user about the possible security concerns. In any case, until 3G networks reach a universal and full implementation level, extra caution and common sense combined with encryption status control (or additional encryption mechanism) via third party applications are strongly suggested.

References

1. OpenBTS, http://openbts.sourceforge.net
2. OpenBSC, http://openbsc.osmocom.org/trac/wiki/OpenBSC
3. Agilent Technologies (1998) 8922M/S GSM Test Set User Guide, Agilent 08922-90211, UK,
4. Racal 6103B Digital Radio Test Set User Manual, Racal Instruments Ltd, UK, 1999
5. Rohde & Schwarz (1999) Digital Radiocommunication Tester CMD52/55, Operating Manual. Germany
6. Qixiang Electron Science & Technology Co. Ltd. (2006) AnyTone AT-400 GSM, Repeater User Manual, China
7. Itu-T E (2008) 212, The international identification plan for public networks and subscriptions, May 2008
8. Marcin Wiacek, Marcin's Page On-line. http://www.mwiacek.com/
9. Digital Cellular Telecommunications System (Phase 2); Mobile Radio Interface Layer 3 Specification (GSM 04.08), Doc. ETS 300 557, 1997.
10. Digital Cellular Telecommunications System (Phase 2+); Radio Subsystem Link Control (GSM 05.08 v. 8.5.0 Release 1999), Doc. ETSI TS 100 911 v. 8.5.0 (2000-10), 1999.
11. Digital Cellular Telecommunications System (Phase 2+); Functions Related to Mobile Station (MS) in Idle Mode and Group Receive Mode (GSM 03.22 v. 8.3.0 Release 1999), Doc. ETSI TS 100 930 v. 8.3.0, (2000-01), 1999
12. Iidakis I, Kandus G (2011) Ramifications of mobile phone advanced O/S on security perceptions and practices. In: Proceedings of the 3rd International Workshop on Cyberspace Safety and Security (CSS2011), September 2011, pp 33–38
13. ETS 300 977 (GSM 11.11 version 5.5.0), European Telecommunications Standards Institute, Digital cellular telecommunications system (Phase2+); Specification of the Subscriber Identity Module Mobile Equipment (SIM ME) interface, May 1997
14. ETS 300 506 (GSM 02.09 V4.4.0), European Telecommunications Standards Institute, Digital cellular telecommunications system (Phase 2); Security aspects, July 1997
15. ETS 300 505 (GSM 02.07 version 4.8.2), European Telecommunications Standards Institute, Digital cellular telecommunications system (Phase 2); Mobile Stations (MS) features, January 1998
16. GSM 11.11 V8.1.0, European Telecommunications Standards Institute, Digital cellular telecommunications system (Phase 2+); Specification of the Subscriber Identity Module Mobile Equipment (SIM ME) interface, November 1999
17. 3GPP TS 31.102 V6.5.0, 3rd Generation Partnership Project; Technical Specification Group Terminals; Characteristics of the USIM application, March 2004
18. 3G TS 22.101 V3.8.0, 3rd Generation Partnership Project; Technical Specification Group Services and System Aspects, Service aspects; Service principles, December 1999
19. 3GPP TS 22.101 V8.11.0, 3rd Generation Partnership Project; Technical Specification Group Services and System Aspects Service aspects; Service principles (Release 8), March 2009
20. Androulidakis I, Pylarinos D, Kandus G (2011) Ciphering Indicator approaches and user awareness. Submitted to MIJST (Maejo International Journal of Science and Technology)

21. GnuRadio, http://gnuradio.org
22. AirProbe, http://svn.berlin.ccc.de/projects/airprobe
23. OpenBTS, http://openbts.sourceforge.net
24. OsmoBTS, openbsc.osmocom.org/trac/wiki/OsmoBTS
25. OpenBSC, http://openbsc.osmocom.org/trac/wiki/OpenBSC
26. OsmocomBB, http://bb.osmocomm.org
27. ETTUS USRP, www.ettus.com
28. HackRF, http://greatscottgadgets.com/hackrf/
29. BladeRF, http://nuand.com/
30. SysmoBTS, http://www.sysmocom.de/products/sysmobts
31. UmTRX, umtrx.org
32. OpenBSC, openbsc.osmocom.org/trac/wiki/osmo-bsc

Chapter 4
Software and Hardware Mobile Phone Tricks

Abstract Mobile phone handsets and networks offer a wealth of applications and functionalities that the average user never gets to use. Nonetheless, they are affecting the security of the phone, either positively or negatively. In this chapter, we will examine a few of them, namely, Net Monitor, GSM/UMTS network codes, mobile phone codes, and the AT command set. There is also a brief software section since, as stated earlier, the software applications and their security fit better in the context of a computer literature work.

Keywords Netmonitor • Engineering menu • AT commands • Network codes • Handset codes • Mobile phone spying software • Mobile phone hardware • Caller ID • Catalog • SMS PDU

4.1 Introduction

Mobile phone handsets and networks offer a wealth of applications and functionalities that the average user never gets to use. Nonetheless, they are affecting the security of the phone, either positively or negatively. The reader should bear in mind that the less he knows about the features, the less protected he is against social engineering attacks. Furthermore, there are options that can help him assess the security level of his device. As such, in this chapter we will try to shed some light on a few of them, namely, Net Monitor, GSM/UMTS network codes, mobile phone codes, and the AT command set. There is also a brief software section since, as stated earlier, the software applications and their security fit better in the context of a computer literature work. We will also see some hardware tricks and modifications. The chapter will end with some hardware tricks and modifications.

4.2 Netmonitor

Netmonitor (or Engineering Menu or Field Test Display [1–4] depending on the manufacturer) which was mentioned previously is a special mode in mobile phones, used to measure network and phone operating parameters and status.

© Springer International Publishing Switzerland 2016   47
I.I. Androulidakis, *Mobile Phone Security and Forensics*,
DOI 10.1007/978-3-319-29742-2_4

Fig. 4.1 Netmonitor
screenshot showing
information about the
serving cell

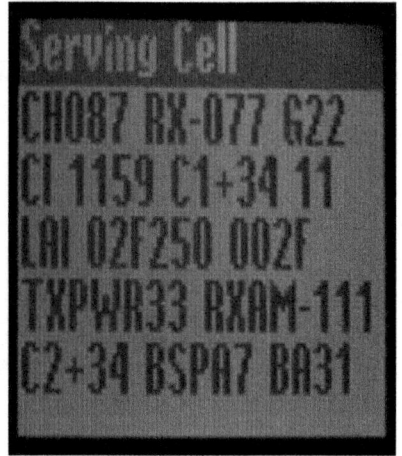

When activated, a new, additional menu usually appears, providing a wealth of information (Fig. 4.1). Dozens of interesting parameters are displayed. To name a few, we get parameters regarding the network, serving and neighboring cell information, the peripherals connected, the SIM status, and the cryptography level, etc. More advanced settings allow the user to actively control the mobile phone. A typical example would be restricting it to certain cells only. It must be noted that each phone has different capabilities in regard to Netmonitor and there are different methods of activation [5, 6]. In older phones, activation was possible by Modifying the EEPROM (the memory holding the firmware) of the phone, or by entering special codes in the keyboard, or by entering specific values in specific places in the SIM catalog or by issuing special commands (called AT commands as we will promptly see). In most cases, a serial port cable and the respective program (such as Gammu [7]) would enable the task. For modern, smart phones, the functionality can be implemented by downloading specific software.

4.3   GSM Network Service Codes

4.3.1   In General

GSM users enjoy services such as diversion—call forwarding, barring, call waiting, answering machine and caller id restriction, PIN changing, etc. Most mobile phones implement these services through menus but there are usually limitations. GSM standards, however, describe the use of specific service codes to enable/setup and take advantage these features, without necessarily using the phone's menu. The interesting point, here, is that by keying the service codes, the user is able to fully control these services since most mobile phones offer a limited only implementation of all the possible characteristics of the services. In addition, providers do not always stick strictly to the standard and implement services differently, leading sometimes to unpredicted situations that can be assessed using these codes.

Most of the codes follow the pattern given in [8] describing the Man-Machine Interface (MMI) of the User Equipment (UE). The structure used consists of these parts:

- Service Code, SC (2 or 3 digits);
- Supplementary Information, SI (variable length).

The procedure always starts with *, #, **, ##, or *# and is finished by #. Each part within the procedure is separated by *. It must also be noted that in most cases, following the codes, the [SEND] key has to be pressed at the end as if they were numbers to be dialed. The [SEND] key is the key the user presses to start a call, labeled in most mobile phones as a small handset picture.

As such, the notion that is used follows below.

Activation	:	*SC*SI#
Deactivation	:	#SC*SI#
Interrogation	:	*#SC*SI#
Registration	:	*SC*SI# and **SC*SI#
Erasure	:	##SC*SI#

Where, according to [9], each action has the following meaning:

Registration: The programming by the service provider or subscriber of information to enable subsequent operation of a service. The programming action involves input of specific supplementary information. For certain services, the registration procedure may cause activation while for others the service may already be in the action phase.

Erasure: The deletion by the service provider, the subscriber, or the system of information stored against a particular service by a previous registration(s).

Activation: An action taken by either the service provider, the subscriber, or the system to enable a process to run as and when required by the service concerned, resulting in the active phase. Some services can be either "operative" or "quiescent" (not operative) during the active phase according to whether or not the system would be able to invoke or use the service.

Deactivation: An action taken by either the service provider, the subscriber, or the system to terminate the process started at the activation.

Interrogation: The request by the subscriber to the mobile network to provide information about a specific supplementary service.

In addition, GSM defines different communication types/services such as voice, fax, data, SMS, etc (known as Bearer services and Teleservices). In [8] again, these services are mapped to MMI (Man Machine Interface) Service Codes that can be used in Supplementary Information field to precisely define to what service the feature will apply to. A selection of these MMI service codes is given in Table 4.1.

Table 4.1 MMI service
codes

All tele and bearer services	No code required
All teleservices	10
Telephony	11
All data teleservices	12
Facsimile services	13
Short Message Services	16
All teleservices except SMS	19
All bearer services	20
All async services	21
All sync services	22
All data circuit sync	24
All data circuit async	25
All GPRS bearer services	99

Table 4.2 Call forwarding of all call types

	Immediate	No Answer	Unreachable	Busy
Set	**21*destination#	**61*destination*nn# nn=delay (5–30 s)	**62*destination#	**67*dest#
Cancel	##21#	##61#	#62#	##67#
Query	*#21#	*#61#	*#62#	*#67#
Cancel All	##002#			

Table 4.3 Call forwarding of voice calls only

	Immediate	No Answer	Unreachable	Busy
Set	**21*destination*11#	**61*destination*nn*11# nn=delay (5–30 s)	**62*dest*11#	**67*dest*11#
Cancel	##21*11#	##61*11#	#62*11#	##67*11#
Query	*#21*11#	*#61*11#	*#62*11#	*#67*11#

It must explicitly be noted that not all telecom providers implement all of these network and MMI service codes, so the user has to experiment or ask the provider for the full applicability of the examples given below.

4.3.2   Forwarding Services

In Table 4.2, we have placed the service codes for call forwarding of all call types, to another number (destination number). Bear in mind that following the codes the [SEND] key has also to be pressed at the end as if they were numbers to be dialed:

To make the point with MMI service codes of Table 4.1 more clear in the next example (Table 4.3), we will forward only incoming voice calls (and not fax or data calls). As you can see, we have added the value 11, in the supplementary information field.

Table 4.4 Barring

	All Outgoing	All Incoming	
Set	**33*barring code#	**35*barring code#	
Cancel	##33*barring code#	##35*barring code#	
Query	*#33#	*#35#	
	International Outgoing	International outgoing except for the home network	Incoming when out of home network (Roaming)
Set	**331*barring code#	**332*barring code#	**351*barring code#
Cancel	##331*barring code#	##332*barring code#	##351*barring code#
Query	*#331#	*#332#	*#351#

Substituting 11 for 25, we get Call forwarding of Data Calls only, while substituting 11 for 13, we get Call forwarding of Fax Calls. According to the standards, SMS too can be forwarded using the MMI code 16, reserved for SMS. A possible string would be **21*destination*16#, requesting the immediate forward of SMSs to "destination" number. Most providers, however, do not implement the feature.

4.3.3   Barring Services

Call Barring allows the user to bar incoming, outgoing, or both types of calls. The user can also bar incoming SMS. For barring of incoming calls, no diversion should be active. Table 4.4 presents some barring combinations.

Again, Using the MMI codes described in Table 4.1, the user can choose the service that is subject to barring. He can, for example, bar incoming SMS using the code: **35*barring code*16#

In order for call barring not to be abused by a third party (removing the restrictions or adding restrictions to the phone without the user knowing it), the functionality is protected by the respective barring code. Most providers initially have set an easy-to-remember value, such as 1234, 0000, or 1111. The user has to ask the operator for the code and then she can change it using the sequence:

*03*330*OLD_PASSWORD* NEW_PASSWORD*NEW_PASSWORD#

The 330 after *03* is used to denote change of password for barring services. If the user wishes to change to a universal password for all supplementary services, she can omit 330.

4.3.4   Passwords

Speaking of code changing, SIM's PIN can also be changed using the following codes. This feature can be abused by social engineering, convincing a user to change his PIN to a number that the fraudster will know beforehand.

**04*oldpin*newpin*newpin# : Change SIM pin code
**042*oldpin*newpin*newpin# : Change SIM pin2 code

Finally, if the SIM is blocked due to many wrong PIN entries, it is also possible to unblock it and enter a new PIN using the PUK as follows

**05*PUK*newpin*newpin# : Unblock SIM pin code
*052*PUK2*newpin*newpin# : Unblock SIM pin code

Since the user has to initially use the present PIN, a social engineering attack could be even easier. Consider this example: You get a call from somebody claiming to be with your network operator. He says that due to maintenance purposes, you have to enter one of the codes we just described. The current PIN is of course needed. To make himself appear as a valid technician of the operator, he will say to you: "I don't want you to tell me your PIN number, this is secret and you should never tell it anybody for your own security. Please enter it by yourself." This way, you might actually think that this is a nice person, while he has just persuaded you to change your SIM's PIN code!

4.3.5 Calling Line and Caller Identification

Identification of the caller to the calling party is one of the handiest features GSM offers. Of course, the ability to withhold the caller ID from being presented to the called party has been associated with an array of unpleasant acts, such as harassments, pranks, spam calls, calls from unwanted persons, and so on.

More specifically, the Calling Line Identification Presentation (CLIP) Supplementary Service provides the called party with the possibility to receive the line identity of the calling party. The network shall deliver the calling line identity to the called party at call set-up time, regardless of the terminal capability to handle the information. The Calling Line Identification Restriction (CLIR), on the other hand, is a Supplementary Service offered to the calling party to prevent presentation of the calling party's line identity, to the called party. These features can be regulated as follows:

CLIR (function of the calling party): This feature can be permanent or temporary. In temporary mode, the user can decide whether to send his caller ID or not to the called party. Please note that the codes given do not have the *XX*destination# form but rather the *XX#destination# form which is somehow different than the norm (the destination number is separated from the command using a # instead of a *)

Send ID: *31# destination_number (recipient sees your number)
Withhold ID: #31# destination_number (recipient does not see your number)
Query: *#31#

The user can also save these codes in the phone's catalog before an entry. So, if the entry #31#6912345678 is saved, I will be calling 6912345678 with caller ID restriction using that catalog entry.

Calling Line Identification Presentation (CLIP, function of the called party):

Allow: *30# (recipient can see the ID of the calling party—unless the latter used CLIR)
Prevent: #30# (recipient does not see the ID of the calling party)
Query: *#30#

As expected, most users would like to have the feature available.

In addition to the caller's ID, there is also the connected line ID. It is very useful in cases where the number called has a forward enabled and as such the ringing phone is not the actual phone that the caller called.

The Connected Line Identification Presentation (COLP) Supplementary Service provides the calling party with the possibility to receive the line identity of the connected party. This Supplementary Service is not a dialing check but an indication to the calling subscriber of the connected line identity. In a full ISDN/GSM environment, the connected line identity shall include all the information necessary to unambiguously identify the connected party. Again, the network shall deliver the connected line identity to the calling party regardless of the terminal capability to handle the information.

In case the connected party wishes to prevent presentation of its line identity to the calling party, she can use the Connected Line Identification Restriction (COLR) Supplementary Service. This could be the case where a business number is forwarded to a personal number and the user wishes to keep his personal number private. It is interesting to note that COLP applies to the caller while COLR applies to the called person. Once again, the respective codes for these features are as follows:

Connected Line Identification Presentation—COLP (function of the calling party):

Activate: *76# (the caller will see the final connected number)
Deactivate: #76# (the caller will not see the final connected number)
Query: *#76#

As an example, suppose number A is forwarded to number B. If caller C has enabled the feature, then when she calls number A and the call gets answered she will read in her screen number B.

Connected Line Identification Restriction—COLR (function of the called party):

Activate: *77# (the caller will not see the final connected number)
Deactivate: #77# (the caller will see the final connected number)
Check: *#77#

Continuing with the same example, if the owner of A does not want C to realize that the call is forwarded to B, he has to use the code for COLR.

4.3.6 Call Management

Among the most interesting features that can be managed with network codes are the call management features that allow call waiting and multiparty conference calling (three way or more) [10]. Sometimes the user has to subscribe to the service, paying an extra monthly fee, while in other operators it is free to use.

Table 4.5 Call management features

0 followed by SEND	Releases all held calls or sets User Determined User Busy (UDUB) for a waiting call
1 followed by SEND	Releases all active calls (if any exist) and accepts the other (held or waiting) call
1X followed by SEND	Releases a specific active call X
2 followed by SEND	Places all active calls (if any exist) on hold and accepts the other (held or waiting) call
2X followed by SEND	Places all active calls on hold except call X with which communication shall be supported
3 followed by SEND	Adds a held call to the conversation
4 followed by SEND	Connects the two calls and disconnects the subscriber from both calls (ECT)
4 * "Directory Number" followed by SEND	Redirect an incoming or a waiting call to the specified directory number
5 followed by SEND	Activates the Completion of Calls to Busy Subscriber Request
Directory Number followed by SEND	Places all active calls (if any exist) on hold and sets up a new call to the specified Directory Number
END	Releases the subscriber from all calls (except a possible waiting call)

Call waiting can be set with *43#, canceled with #43# and its status is checked with *#43#. After that, the user will be able to receive a second call while talking. When a call is waiting to be answered, the following actions can be taken by pressing the respective keys, as seen in Table 4.5. "X" is the order (starting with 1) of the call given by the sequence of setting up or receiving the calls (active, held, or waiting) as seen by the served subscriber. Calls hold their number until they are released while new calls take the lowest available number.

Using the previous codes, it is possible to start a three way calling teleconference where three callers are connected and can speak simultaneously. Again, depending on the provider, a monthly fee might be needed to enable the service. This is the order of commands to achieve that:

1. Call A
2. Call B (A goes into hold as we described right before)
3. Press 3 [SEND] and you can now speak all together
4. You can drop A or B out of the teleconference by pressing 1 or 2 and [SEND].

The feature works for two incoming calls too, not necessarily two outgoing.

The other two parties (excluding the party that started the teleconference) are getting an informatory beep in regular time intervals (e.g. every 15 s). This is a very important point. As we will see later, there exist various "spy" software tools that are abusing the three way calling feature. Indeed, what they do is that, when the victim is talking on the phone with somebody, without the user realizing, they are starting a three way calling conference with the number of the attacker. So the attacker "enters" in the conversation and can listen to it. As we said, periodical "beeps" will be heard in the two callees (not to the victim's phone however, since it appears that she has originated the service!). The attacker of course knows what this tone is. So, the only one that can realize that something is wrong is the initial party that the victim talked with. Needless to say that very few people realize that this tone is actually a warning about the presence of another party in the conversation.

4.4   Mobile Phone Codes

Apart from the GSM network codes, handsets have their own short codes implemented, most of them not documented. Using these codes, the users can achieve various tasks [11]. They can be simple such as resetting the phone or changing the Language to the default one (after e.g. having accidently set it to Greek!). More advanced tasks include unlocking the phone to be used with SIM cards from other providers, enabling special menus and features such as Engineering Menu (or Netmonitor as discussed earlier), or validating the age of the phone and the warranty status. It can also lead to test procedures to test the functionality of the phone and its peripherals. It is interesting to note that such phone codes have found their way to legendary-hoax emails. This is a typical example with the code that toggles between Full Rate/Half Rate voice codec. Mobile phones can use a higher quality and a lower quality codec to transmit the voice to the network, known as full rate and half rate, respectively. When using the half rate, the quality of the communication is lower but the power consumption is also lower. As such there is gain in the battery life. This feature was presented in an email circulating the Internet as a secret code that "magically" enhances the talk time of mobile phones. Of course it was nothing of a "magic" but rather the option we just described.

One of the most known codes (or at least one of the codes that should be among the most known) is *#06#. This is the default code to enter in GSM/UMTS handsets to display the IMEI [12] (International Mobile Equipment Identification code). The IMEI is used to identify the mobile equipment (ME) or the user equipment (UE) as it is named in UMTS networks. Each ME has a unique IMEI. This is a unique 14-digit number in each cell phone as depicted in Fig. 4.2. According to [8], the procedure shall be accepted and performed with and without an inserted SIM. Indeed, in Fig. 4.3, the IMEI is presented even without a SIM card. The ME displays the 14 digits of the IMEI, the check digit and optionally the software version number as defined in [12] (as a single string, in that order). The IMEI is hard-coded in the ME

Fig. 4.2 IMEI as displayed in the screen of the phone

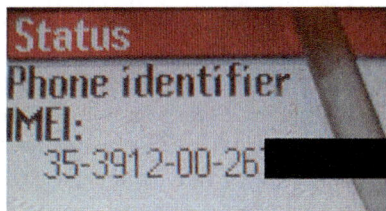

Fig. 4.3 IMEI presented
even without a SIM
inserted

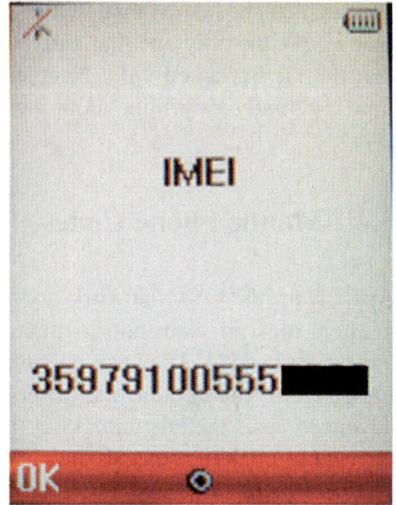

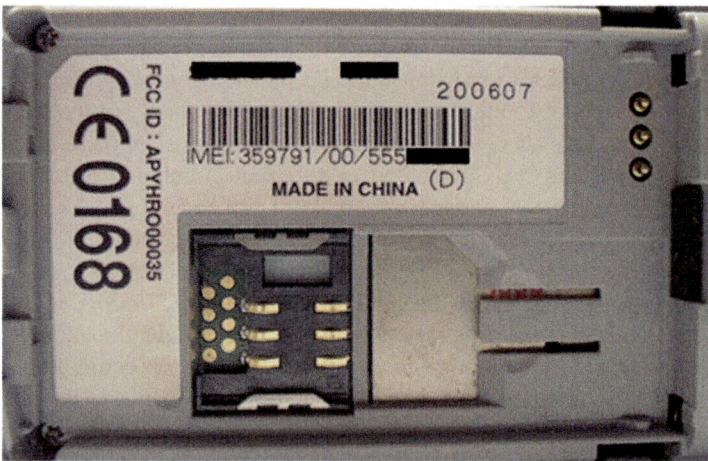

Fig. 4.4 IMEI written in the sticker beneath the battery

and it is not used for routing or subscriber identification. Operators can in theory
block a cell phone whose IMEI number is registered in a "black list" (e.g. because it
is stolen). This register is known as the EIR (equipment identity register). In any case,
the user should note down his IMEI for possible future reference. The same information is written usually beneath the battery of the phone, as portrayed in Fig. 4.4.

Continuing, as an example, we are adding an assortment of different codes used
in different phones. The reader is advised not to use these codes in his primary
phone but rather in a cheap test phone. Fig. 4.5 presents the result of one of these
codes entered in the phone.

Fig. 4.5 Software version,
given by pressing the
corresponding code

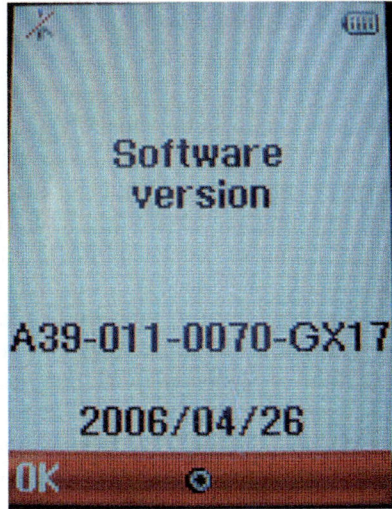

Manufacturer A

*#0000#—software version
#7780# (#RST0#)—reset phone
#92702689# (#WAR0ANTY#)—warranty status
*#WARRANTY#—warranty status
*#WAR0ANTY#—warranty status
#4720# (#HRC0#)—enable half rate codec
#3370# (#EFR0#)—enable full rate codec

Manufacturer B
Right_arrow*Left_arrowLeft_arrow*Left_arrow*—Secret Menu
Up*DownDown*Down*—Service Menu
Left**Left—Lock status
Left0000Right—Language reset

Manufacturer C
*01763*3640# (*01763*ENG0#) --- Disable Engineer Mode
*01763*3641# (*01763*ENG1#) --- Enable Engineer Mode
*01763*4634# (*01763*IMEI#) --- IMEI
*01763*63866330# (*01763*NETMODE0#) --- Disable NetMode
*01763*63866331# (*01763*NETMODE1#) --- Enable NetMode
*01763*737381# (*01763*RESET1#) --- System reset (all default set-tings)
*01763*79837# (*01763*SWVER#) --- Software version
*01763*8371# (*01763*VER0#) --- Firmware Version

4.5 Catalog Tricks

Since we are focusing on mobile phone codes and features in this section, we will focus on catalog's functionality. On most phones, a shortcut to last dialed numbers can be achieved by 0#. For a shortcut to numbers stored in SIM's catalog, the user types on main menu a number (order of the entry in the SIM catalog) followed by #. As an example, 10# would present the tenth number stored in the SIM. In cases where a number is deleted from the catalog, the empty slot can be found and as such somebody can deduce that a number has been erased from that position.

Another "feature" that can be manipulated by fraudsters trying to social engineer a user is directly connected to the catalog's functionality. It is relevant to the way mobile phones map the numbers dialed (or the incoming numbers calling) to the respective entries in their catalog list. The catalog application in most cell phones matches only the 6–8 last digits of a number in order to show the relative entry name. As an example, let's assume that the number +306912345678 corresponds to the entry "Iosif"in the catalog. This entry includes the country prefix (+30 for Greece) and the national mobile phone number 6912345678 (10-digit Greek national number planning). Now, if the user dials just 345678 (or 2345678, or 12345678, depending on the phone) and hits the "dial" button, the phone will try to dial that number. Of course it is not a valid number so the call will not proceed. The interesting part is that during the call setup, the display will show "Iosif," having performed the number matching using only the last digits and not the whole number.

No harm done, so far. But, consider the following scenario, where the user is dialing 2651008888345678. As stated earlier, the Greek national numbering plan is 10 digits. So the number that the phone will dial is 2651008888 (the first 10 digits used, the rest is discarded). However, as we just described, the phone's catalog application will match only the 6–8 last digits of the dialed string. So the display will show "Iosif" (matching the ending of the string 345678, since Iosif's number is 6912345678). The phone will be calling a certain party but the display will be showing a different contact! A fraudster, using social engineering techniques, could persuade a user to dial such a long sequence, exploiting various fraud schemes.

Why is this "feature" not corrected? The answer has to do with roaming and the use or not of country codes in the caller id. Indeed, if the catalog application demanded the full number to completely match then in case caller id had (or had omitted) the country code, the match would not have been possible. Using the initial example, if the local number 6912345678 was entered in the catalog exactly as that, then if caller ID presented the number as +306912345678, the catalog would not be able to match it to the contact (since the initial +30 would be missing). Or, the other way around, if the user had entered the number as +306912345678, then calling 6912345678 and not +306912345678 would not show the name entry.

4.6   AT Command Set

4.6.1   In General

Leaving aside the phone codes, we will now enter a more technical field, dealing with "AT commands." The use of AT commands (AT stemming from "Attention") is a very useful way to set up and control modems. Extended command sets can also control other communication devices too. It was initially developed for the Hayes Smartmodem in 1981 but following its universal acceptance, the command set was also adopted for mobile phones, as defined in GSM 07.07 [13].

In this command set, there are different commands for

* Identification (IMEI, IMSI, Manufacturer, etc.)
* Call Control (Place a call, hang up the phone, call forward, auto answer)
* Device Control (switch it on/off, place keystrokes)
* Catalogs (full access read/write)
* Messages (full access read/write/send)
* Miscellaneous other functions

Apart from the standardized set of commands, manufacturers are also using their own supersets adding platform specific commands [14]. All these commands are issued to the mobile phones' processor using a serial port cable (Fig. 4.6) and a terminal communications program such as HyperTerminal. Below, there is an assortment of these codes, grouped by their functionality.

Fig. 4.6 Serial port cable for connecting the phone to PC and issuing of AT commands

4.6.2   Identification

- AT+CLAC list all available commands
- AT+CGMI, manufacturer
- AT+CGMM, model
- AT+CGMR, revision
- AT+GMI, manufacturer
- AT+GMM, model
- AT+GMR, revision
- AT+CGSN, IMEI (same as *#06#)—example in Fig. 4.7
- AT+CIMI, IMSI—example in Fig. 4.7
- ATI, identification
- AT+CNUM, number
- AT+CCLK?, date and time—example in Fig. 4.8

4.6.3   Call Control

- AT+CHUP, hung up
- AT+CREG, net registration
- AT+COPN operator names
- ATD, dial

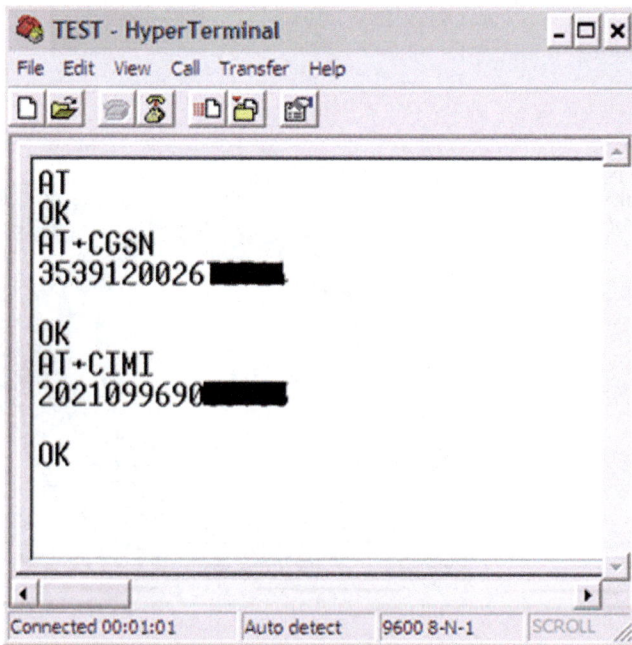

Fig. 4.7 Results of identification commands

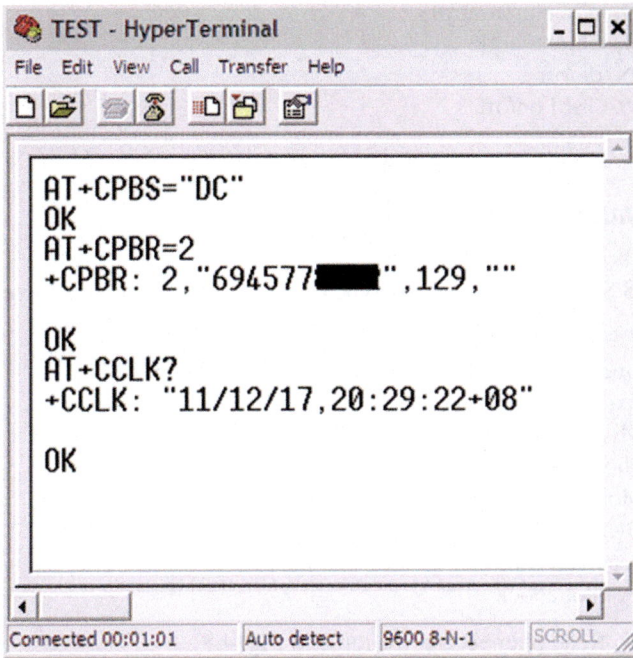

Fig. 4.8 Dialed call and clock

- AT+CLIP, caller id
- AT+CPAS, state of mobile
- AT+CBC, battery
- AT+CSQ, signal quality
- AT+CACM, call meter
- AT+CRMP, playback melody
- AT+VTS, send DTMF
- AT+CCFC, call forward
- ATS0 = 1, autoanswer

ATD is one of the most useful commands. Entering ATD 6912345678, the mobile phone will dial number 6912345678 and the display of the phone will show the process, the same way as when dialing normally using the keypad. The ATD sequence can also be used to issue network codes. Issuing ATD *21*destination number# would enable a call forward of voice calls to the destination number.

4.6.4   Device Control

- AT+CKPD, keypad control
- AT+CMER, event reporting
- AT+CKEV, event key

- AT+CIND, indicator control
- AT+CVIB, vibrate mode
- AT+CFUN, on/off
- AT+WS46, GSM on/off

4.6.5   Catalogs

- AT+CPBS Select the phonebook type among the following (example in Fig. 4.8)

  <nl>"DC"—Dialed calls
  "EN"—Emergency numbers, writeprotected stored on SIM
  "FD"—Fixdialing numbers
  "MC"—Missed calls
  "ME"—Phonebook
  "MT"—Mobile memory + SIM phonebook
  "ON"—Own number
  "RC"—Received calls
  "TA"—TA phonebook

- AT+CPBR, Read phonebook (example in Fig. 4.8)
- AT+CPBR = ? Number of supported entries from the phonebook
- AT+CPBR = 7 Display entry number 7 of the phonebook
- AT+CPBF Search by name in phonebook

  AT+CPBF = "Iosif" (Searches for an entry with name "Iosif" and displays it)

4.6.6   SMS

SMS control via AT is governed by GSM 07.05 [15] as follows:

- AT+CPMS = "SM", select SIM to read SMSs
- AT+CPMS = "ME", select mobile phone's memory to read SMSs
- AT+CMGL = 0, list received unread messages
- AT+CMGL = 1, list received read messages
- AT+CMGL = 2, list stored unsent messages
- AT+CMGL = 3, list stored sent messages
- AT+CMGL = 4, list all messages (example in Fig. 4.9)
- AT+CMGR, read message
- AT+CMGR = n, read SMS with index n
- AT+CMGS, send message
- AT+CMGW, write SMS to memory

At this point, we will present the way to send an SMS using AT commands, and specifically the AT+CMGS command. We will not cover all the details, but rather we will show a bare bones SMS. The interested reader can find all of the details in [16].

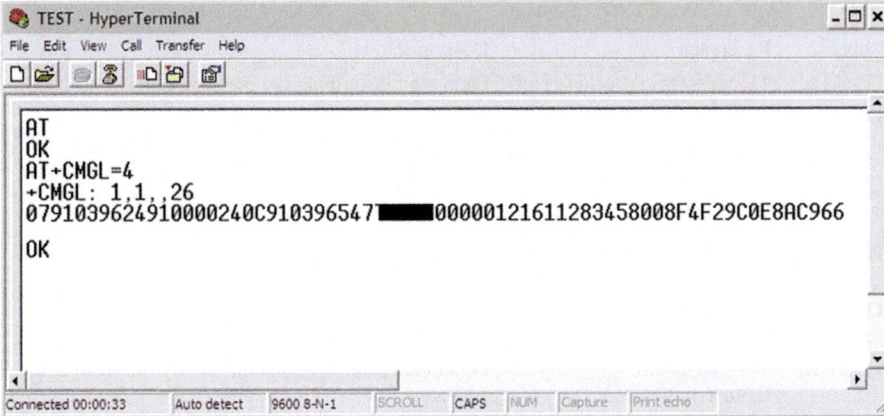

Fig. 4.9 A message read

Information that is delivered as a unit among peer entities of a network (such as in the short messaging service) is presented in Protocol Data Units (PDU). There are different forms of PDUs for SMSs, such as the SMS-SUBMIT (for messages that are sent from the mobile phone), the SMS-DELIVERY (for messages delivered to the phone), and others. In our example, we will focus on SMS-SUBMIT only. An SMS-SUBMIT PDU consists of the Serving Center Address (SCA) and the Transport Protocol Data Unit (TPDU). The SCA is the number of the SMSC that handles the messaging service for the user. An SMS to be sent is formed as SCA+TPDU. Since in most cases the terminal will use the SCA address stored in the SIM card, the first PDU is set to 00. So the SMS becomes 00+TPDU. The exact elements that constitute SMS SUBMIT TPDU can be seen in Table 4.6. It must be noted that not all of them are mandatory. Some of them are only 1 bit long while others are many bytes long (i.e. the User Data Part)

The command AT+CMGS is interactive. That is, the user initially enters the command with the length of the data packet in bytes (in decimal) that will follow (excluding the length of the SCA field) and presses enter. Bytes consisting of 8 bits are also called octets. As an example, consider the bare bones SMS:

AT+CMGS = 14 //Send message, with a TPDU length of 14 octets (excluding the
    initial byte 00 which is part of SCA)

The phone replies with the prompt ">" and the user has to enter the TPDU field in hexadecimal form. Then he has to press CTRL-Z (denoted as ^Z) to end the sequence and the message will be send.

>0021000c9103961395827700C801D4^Z
Let's now split this into its parts
00 21 00 0c 91 039613958277 00 C8 01 54

The first octet ("00") (that is not counted in the length submitted to AT+CMGS since it is part of SCA) denotes that the handset will use the SMSC information stored in the SIM.

Table 4.6 SMS-SUBMIT TPDU elements

Abbr.	Reference	Description
TP-MTI	TP-Message-Type-Indicator	Parameter describing the message type
TP-RD	TP-Reject-Duplicates	Parameter indicating whether or not the SC shall accept an SMS-SUBMIT for a Short Message still held in the SC which has the same TP-MR and the same TP-DA as a previously submitted SM from the same Originating Address
TP-VPF	TP-Validity-Period-Format	Parameter indicating whether or not the TP-VP field is present
TP-SRR	TP-Status-Report-Request	Parameter indicating if the MS is requesting a status report
TP-UDHI	TP-User-Data-Header-Indicator	Parameter indicating that the TP-UD field contains a Header
TP-RP	TP-Reply-Path	Parameter indicating the request for Reply Path
TP-MR	TP-Message-Reference	Parameter identifying the SMS-SUBMIT
TP-DA	TP-Destination-Address	Address of the destination entity
TP-PID	TP-Protocol-Identifier	Parameter identifying the above layer protocol, if any
TP-DCS	TP-Data-Coding-Scheme	Parameter identifying the coding scheme within the TP-User-Data
TP-VP	TP-Validity-Period	Parameter identifying the time from where the message is no longer valid
TP-UDL	TP-User-Data-Length	Parameter indicating the length of the TP-User-Data field to follow
TP-UD	TP-User-Data	The Actual Message that will be send

The next octet (21) is a binary mapped field that contains the six first elements of Table 4.6. If we name the 8 bits of it as b7,b6,b5,b4,b3,b2,b1,b0, then the exact mapping is this:

b7 = TP-RP
b6 = TP-UDHI
b5 = TP-SRR
b3b4 = TP-VPF
b2 = TP-RD
b0b1 = TP-MTI

Transforming hexadecimal 21 that we used in our example to binary, we get 00100001. This translates as TP-MTI=01 (SMS-SUBMIT), TP-RD=0 (do not reject duplicates), TP-VPF=00 (do not use validity period), TP-SRR=1 (ask for a status report delivery), TP-UDHI=0 (no special data header in User Data part), TP-RP=0 (no reply path needed). We are therefore sending an SMS.

Then, there is the Reference parameter, which is handled by the mobile handset, incrementing it serially for each new message. As such, it is sufficient to leave the value 00.

Following, 0C which translates to decimal 12, is the length of the Destination Address number in digits. Before the number itself, another octet denotes the type of the addressing used. In most cases, hexadecimal 91 is used which means that the recipient's address is coded with the ITU-T E164 [17] format (the well-known international country prefixes).

After that, the actual number comes, with a "twist" however. The digits are presented in swapped couples. To make this clear, this was our example: 039613958277. The couples are 03 96 13 95 82 77. We have to swap them so we get 30 69 31 59 28 77, which is a Greek mobile network number +306931592877.

TP-PID, coming next, is a very important part, since it indicates the protocol functionality of the message. The normal value for a simple SMS is 00. But there are far more interesting options. Value 40 sends a ping SMS, technically called short message type 0. This type of message indicates that the mobile equipment must acknowledge its receipt but shall discard its contents without storing them. This is why it can be used to "ping" the mobile phone to see if it is switched on or off. Value 7F sends a binary SMS to be directly downloaded to the SIM of the destination number. Values 41–47 replace an already stored SMS in the destination that happens to have the same value and has been sent from the same originating number. An example of this will be seen on Sect. 5.5.

Since TP-DCS has so many options that is devoted a whole standard by itself [18]. Among the most interesting values are the class 0 message (also known as Flash message) and the message indicator control values. Class 0 messages appear immediately on the screen of the phone, without the user performing the normal procedure to read them as with normal messages. The respective TP-DCS value is 10 or F0 (2 different values with the same result), while their possible malicious use will be analyzed in Sect. 5.5.

Continuing with TP-DCS, in mobile phones there are some specific icons-indicators to alert the user for the presence of a voicemail, a fax, an email, or other kind of message waiting. One of the ways to operate these indicators is by setting the appropriate TP-DCS value. The message itself can contain text, e.g., "You have a new voice message" that will be saved as an SMS, or it can be automatically discarded after enabling (or disabling) the respective indicator. As such, TP-DCS values C8,C9,CA,CB activate the voicemail, fax, email and other type indicators without saving the SMS in the mobile phone, while C0,C1,C2 and C3 values deactivate the same indicators. Using D0-D3 and D8-DB instead of C0-C3 and C8-CB, the user can also accompany the operation with a text message that will be stored in the mobile phone. In our example, we have TP-DCS=C8 which means that the voice mail indicator should be lit and the message be discarded.

The next byte would be reserved for the validity period (how long the message will wait stored in the service center if the destination phone is switched off), but in our example we have not used it (see the first byte of the TPDU), so we can skip it.

Closing we have the User Data Length field, where we have to provide the length of the actual User Data field. User data itself is encoded according to TP-DCS (7bit, 8 bit, 16bit alphabets etc). It must be noted that the default value is the 7 bit GSM alphabet. In this case the characters of the message itself are 7 bits each, while each byte is 8 bits. This way, 8 characters can fit in 7 bytes, stealing 1 bit from each byte for the next character and so on. In our example, we are sending a single "T" character so its value is the same with the 8 bit ASCII value.

Once again, the example was:

AT+CMGS = 14
>0021000c9103961395827700C80154^Z

that will "silently," without being stored as an SMS, lit the voicemail waiting indicator. The receiving user would be very much puzzled now because she will not be able to switch the indicator off since she actually does not have any messages waiting in her operator's system.

4.6.7   SIM Access

Closing, two very powerful commands enable low level access to the SIM, reading its contents. We will see more details in the last chapter, dealing with forensics.

*   AT+CRSM = Restricted SIM access
*   AT+CSIM = Generic SIM access

4.7   Software

As previously stated, modern cell phones can download and execute programs the same way computers do. There are millions of applications, games, and utilities available for every platform. At the same time, worries about malicious software increase. Older variants of mobile phone viruses were difficult to be transferred, with Bluetooth being the main vector. It is now easier than ever to infect the target, since the user himself will download the malicious code from the Internet. We won't write much about software and viruses since it is a subject more closely fitting to computer literature [19–21].

We will, however, focus on a special category of software available for mobile phones. It is that of intercepting suites, available for most smart phones today. Indeed, searching in the internet with the terms mobile phone spy will reveal more than half a dozen of such programs [22].

These programs can relay the incoming and outgoing SMSs of the user, the calling history, the sites he browsed, and so on. The position of the phone can also be determined. Using the three way calling feature as described earlier, they can relay voice communication too, enabling the attacker to listen-in to ongoing calls. In addition, they can monitor the surroundings of the user, by silently enabling the microphone. This way they transform to a very effective bugging device.

There are two ways they operate. One is with a client-server architecture, where the software uploads the information to a server and the attacker, using the internet and his account logs in and examines all the intercepted data. The other one relays directly the information using SMSs. The second way of operation is easier to spot in the bill, where SMSs to unknown numbers will appear. All it takes is a couple of minutes access to the phone and a malicious user can "tap" the phone of her target (luckily the installation can't be performed remotely). Using social engineering techniques, she can also convince the victim to install such malicious software, masquerading as an innocent game.

On the other side, it is possible to install in the phone encryption software. Not only stored information but also voice calls can be encrypted. Voice encrypting products work by transforming voice to data and encrypting the data before leaving the phone. As such, the encryption is not depending any more on the provider but rather on the specific software and/or hardware suite installed in the phone [7, 23]. One drawback with such solutions is that the user is charged for data exchange rather than voice service. Therefore she can't benefit from "free minutes" or unlimited talk plans provided by the telecom operators. Needless to say, the recipient of the call must have a compatible product in his phone too.

4.8 Hardware

Before the advent of smart phones, and the wealth of software available, the combination of a mobile phone and a microcontroller was an easy way to provide telemetry applications and to remotely control devices. AT commands were issued to the mobile phone via the microcontroller and the user could remotely interact with the phone. The microcontroller is no more since smart phones can execute code by themselves now. However, it might still be cheaper to use a bare bones mobile phone and a microcontroller (Fig. 4.10) for telemetry applications rather than an expensive smart phone.

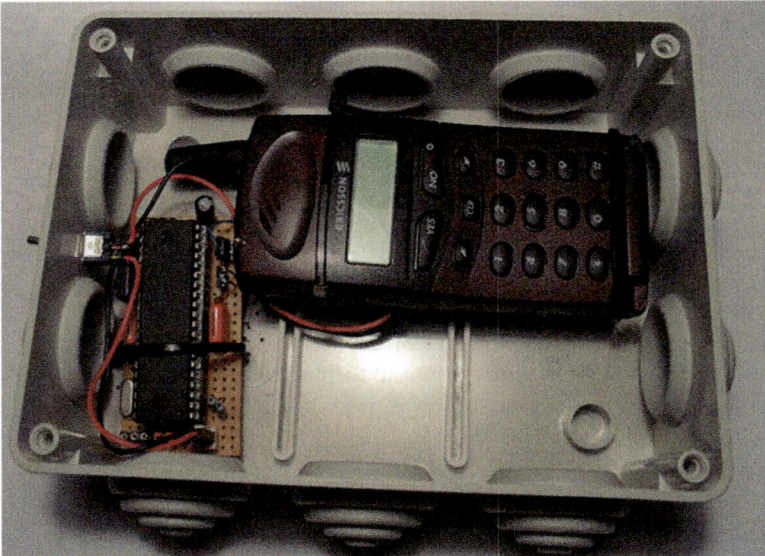

Fig. 4.10 Mobile phone and microcontroller

Once again, before the era of smartphones, hardware modifications were used to achieve various tricks. The most known one is the trick of the forgotten-left behind cell phone. It is a very effective bugging device. Such a phone appears completely dead, but, it is working secretly. When called by the attacker, it switches the microphone on (without ringing or flashing the screen at all) and allows him to monitor the place (worldwide coverage bug!). A businessman would "forget' his phone in the office of a competiting company or a husband would hide such a phone back home and monitor what is going on. More elaborate modifications allow mobile phones to operate normally as every other innocent phone and only start their silent spying when they get a call from a special predefined number. Finally, the combination of GSM and GPS (either with separate hardware modules or with embedded GPS) opens the door to the ability to monitor the whereabouts of the user.

Nowadays, hardware modifications are not needed anymore since smartphones allow programmers to fully take advantage of most features. In a more serious context, however, hardware modifications of mobile phones have been used to detonate bombs and other terrorist actions.

Closing the hardware section, jammers or signal blockers pose significant threats to the availability of the mobile phone service. There exist various models with different ranges of coverage. They can cover from a couple of square meters to a whole city. Their principle of operation is quite simple. They are transmitting in high power in the whole frequency band of GSM. As such they mask legitimate signals and mobile phones cannot operate. Given the simplicity of their goal (to fill the radio frequencies with noise), their construction is relatively easy and their price is low. More advanced models are able to interfere only with the specific signaling channels and not the whole band, selectively jamming communications. It is also possible to use a fake base station, as described in the previous chapter, to effectively block mobile phone communication.

4.9   Conclusion

This chapter has probably been the most diverse themed chapter of the book, covering only a fraction of the wealth of applications and functionalities that the average user of mobile phones never or very seldom gets to use. The reader should bear in mind that the less he knows about the features, the less protected he is against social engineering attacks. Furthermore, there are options that can help him assess the security level of his device. As such, we tried to shed some light on a few of the most interesting ones, namely, Net Monitor, GSM/UMTS network codes, mobile phones codes, and the AT command set. The software section was deliberately short, while in the hardware section we briefly described some hardware tricks and modifications affecting the security of the user.

References

1. Jokinen JP (2004) Field Test Display Specification, DTS08337 EN-1.0, Nokia
2. Quirke J (2004) JQ's Nokia Net Monitor Guide
3. Nokia (2002) User guide for network monitoring menu
4. Nobbi, Nokia NetMonitor Manual, www.nobbi.com
5. Marcin Wiacek, Marcin's Page On-line, http://www.mwiacek.com/
6. Gammu, http://www.gammu.org
7. GSMK Cryptophone, www.cryptophone.de
8. 3GPP TS 02.30 (2002) 3rd Generation Partnership Project, Technical specification group services and system aspects; Man-Machine Interface (MMI) of the Mobile Station (MS)
9. 3GPP TS 02.04 (1998) 3rd Generation Partnership Project, Digital cellular telecommunications system (Phase 2+); General on supplementary services
10. 3GPP TS 02.84, 3rd Generation Partnership Project; Technical Specification Group Services and System Aspects; MultiParty (MPTY) Supplementary Services—Stage, 2005
11. Mobile Zone: Secret Codes, http://twilightwap.com/mobile/secrets.asp
12. 3GPP TS 02.16, 3rd Generation Partnership Project, Technical Specification Group Services and System Aspects; International Mobile station Equipment Identities (IMEI), 2000
13. 3GPP TS 07.07, 3rd Generation Partnership Project, Technical Specification Group Terminals; AT command set for GSM Mobile Equipment (ME), 2003
14. Siemens, AT command set for S45 Siemens mobile phones, 2001
15. 3GPP TS 07.05, 3rd Generation Partnership Project, Technical Specification Group Core Network and Terminals; Use of data terminal equipment Data Circuit terminating Equipment (DTE—DCE) interface for Short Message Service (SMS) and Cell Broadcast Service (CBS)
16. 3GPP TS 03.40, 3rd Generation Partnership Project, Digital cellular telecommunications system (Phase 2+); Technical realization of the Short Message Service (SMS) Point-to-Point (PP)
17. ITU-T E.164: Series E: Overall network operation, telephone service, service operation and human factors; International operation—Numbering plan of the international telephone service; The international public telecommunication numbering plan
18. 3GPP TS 03.38, 3rd Generation Partnership Project, Technical Specification Group Core Network and Terminals; Alphabets and language-specific information
19. Bose A, Shin KG (2006) On Mobile viruses exploiting messaging and bluetooth services. The University of Michigan, Ann Arbor
20. Haas P (2005) Cellular Phone Viruses
21. Amit Kumar Jain, Mobile Viruses and Worms, 2006
22. The Spyphone Guy, http://www.spyphoneguy.com/
23. Caspertech, http://www.caspertech.com/

Chapter 5
SMS Security Issues

Abstract Short Messaging Service is one of the most widely used services of mobile telephony. As we will see in this chapter, there are threats to its Confidentiality, Integrity, and Availability. Even worse, the advent of more advanced capabilities and services, including mobile shopping and mobile banking transactions, which largely rely on the ability to send and receive short text messages to authenticate the user, will raise even stronger security concerns.

Keywords SMS security • SMS confidentiality • SMS integrity • SMS availability • SMS UDH • SMS spamming • Silent SMS • Ping SMS • Flash SMS • Bulk SMS

5.1 Introduction

The first short text message was sent in Britain in 1992, wishing "Merry Christmas" [1]. Despite the fact that mobile telephony was designed and launched commercially for voice transmission over the coming years, the use of short message service escalated rapidly. Throughout its almost 20 years of life, SMS was established as one of the most widely used services of mobile telephony. Typically, only in USA, in 2008, more than 1 trillion SMS [2] were sent bringing gains of several billions to the providers.

SMS operation is based on a store and forward service with which messages received from the mobile user, are stored in a central server-message center, and from there forwarded to the mobile recipient. Storage is necessary in order for the message to be eventually sent if at the time it is sent from the sender, the recipient's phone is switched off or out of coverage. Messages are transferred to the nodes of the providers in clear text according to different formats depending on the manufacturer. The most known ones are SMPP, EMI/UCP, TAP, and others.

The actual way of operation and the specifications are described in detail in the GSM standards [3]. The short message center (SMC—Short Message Center) undertakes to receive, store and send messages from an entity of sending/receiving text messages (SME—Short Message Entity) which may be a mobile telephone, a computer, or some other service network. Then the gateway short message service (GMSC—SMS gateway MSC) which is a specially modified switch, acts as an interface message center with the rest of its network and other providers.

© Springer International Publishing Switzerland 2016

I.I. Androulidakis, *Mobile Phone Security and Forensics*,
DOI 10.1007/978-3-319-29742-2_5

On receiving the short message from the short message center, GMSC uses the SS7 network to interrogate the current position of the mobile station from the HLR (home location register) as to accordingly forward the message.

In the coming pages, we will examine Availability, Confidentiality, and Integrity issues involving the SMS. It must be noted that the advent of more advanced capabilities and services, including use of electronic shopping using mobile phones (m-commerce) or banking transactions, which largely rely on the ability to send and receive short text messages to authenticate the user, will raise even stronger security questions [4, 5].

As usually, using a PC helps construct special SMSs that achieve some of the attacks described in the next paragraphs. The way to send SMS via the PC was described in the previous chapter, discussing AT commands.

5.2   Availability Issues

Starting our analysis based on the characteristic of Availability, we observe a series of attacks aiming at causing denial of service. In the simplest scenario, the malicious user sends hundreds or even thousands of messages to the victim's mobile phone. This is possible using either software or more easily by using SMS bulk services from the Internet. If it happens to be an older device, its memory will quickly fill up and thus it will not be able to accept any more messages. Even if the user deletes these messages, more messages that are waiting, stored at the message center, will arrive as soon as the user frees up space. This process can be repeated countless times, effectively blocking the service.

If the attack is launched against a more modern device that has a larger memory capacity, the problem will manifest in a different way. The user will receive messages but will not be able to distinguish immediately the original message intended for him, without having to read it first, within the "sea" of messages sent by the attacker. In addition, the latter could constantly change the appearing sender's number (possibly using numbers belonging to persons that could be in the catalog of the victim) making the situation even more difficult for the recipient. Imagine for a minute having 100 messages waiting, each one from a different number. How do you know which one is valid and which one is not? Furthermore, this continuous flow of messages may even lead some devices to crash and stop functioning.

Respectively, depending on how carefully the software and operating system of the phone is designed, the attacker may be able to send messages that cause "confusion" to the device and also lead to its blockage until they are manually deleted. In some cases, it is also possible to get the phone in a state where it crashes again even if it is rebooted. It might be possible that the offending messages cannot be deleted, leading to a permanent situation. These messages usually contain special characters, long names, invalid characters, or deliberately fail to follow the specifications and requirements of the standards in regard to their structure and syntax, causing an "erratic" response from the device [6–9]. Especially, crafted vcards (virtual business cards that the mobile phones can exchange in lieu of the classical business cards), binary messages with various combinations of TP-PID and DCS, with a broken User Data Header part and other combinations are especially effective. It is also possible for some combinations to lead to a "silent" denial of service, where the phone operates normally but only SMSs fail to be received [10].

On a second level, continuous messaging (especially if messages are "invisible" to the user) can drain the available battery energy of the mobile, causing a denial of service not only for the message service but also for voice service and the device itself, as it was seen in Chap. 1. Furthermore, installing a fake base station (as described in Chap. 3) or even simpler, a jamming device would deprive the mobile phones nearby from the ability to send and receive short text messages. Of particular interest is scientific research [11, 12] which has been published that demonstrates that a relatively small but constant flow of messages may overload and disable even the message sending centers (SMSC) causing massive interruptions to the service.

Of course, a denial of service is not necessarily always due to an attack. A limited, usual problem with sending messages is observed during the holidays where the networks are called to serve millions of SMS within a very short time, leading to overload and delays.

5.3 Confidentiality Issues

As mentioned in previous chapters, encryption (if applicable) for voice transmission as well as for text messages usually stops when the communication arrives at the provider's internal network. From that point on, messages are forwarded within the network without any encryption according to the specifications of the different protocols depending on the manufacturer. Thus, employees of the provider who have access to the corresponding systems can read at will the contents of the messages as well as the sender's and recipient's personal information.

The attacker does not necessarily need to have access to the internals of the network. A few minutes of access to a mobile phone is all what is needed in order to install malicious software that is able to relay to a third number the messages sent and received by this device. Of course, the messages that will be retransmitted by the victim's mobile phone to the mobile phone of the offender are billed normally. So, this activity and the number used to receive the messages will be revealed by a careful examination of the next billing statement. But by then it will be too late…

The fake base station described in Chap. 3 has the ability to intercept any message that the user tries to send as well as the number of the recipient. As we have seen, the fake base station is not part of the actual home network of the victim, so the intercepted message will never reach its destination. But, since the attacker knows the content and the recipient, he can send a copy of the original message to the recipient, using techniques to masquerade and change his/her identity to match the number of the intercepted phone. This is easily accomplished with Internet SMS gateways and as such the attack can succeed without being noticed. We will give more details about SMS masquerading in the next section.

Short messages, in any case, apart from the content itself can reveal information and user behavior violating her privacy. In fact, through the delivery report of messages, a simple user (having no access at all to insider's information) can be informed about whether the mobile phone of another user is switched off and exactly the time the user turns it on, and vice versa. This is possible since the delivery report message will be delivered to the originator as soon as the mobile phone is switched on and will be pending for as long as the mobile phone of the recipient is switched off.

Apparently, sending plain messages (even empty) for this purpose will be immediately spotted. Smart techniques [13, 14] however make it possible to implement this attack without the unsuspecting user ever knowing he is receiving these messages, as it was shown in Chap. 1. It is possible to send "invisible" messages that get delivered normally to the victim's device, but ultimately do not appear on the screen and no notification is given. Such messages can be sent by a simple user not having access to the internals of the network. At the same time, recording the times of delivery reports allows the attacker to "monitor" the behavior of the owner of the mobile phone. Using the previous techniques for some time, the attacker can extract the typical usage pattern related to the target and get in position to know each deviation from it. He could know,for example, that a given day the "victim" overslept (switched on the mobile phone later than usual), or that is still awake (since the mobile phone is still switched on, while the patterning behavior indicates that the user normally switches the phone off at night). In Fig. 5.1, stealth SMSs are sent every half an hour. Delivery reports are noted with blue dots. The red line denotes the time that elapsed from the last delivery report and gives an estimation of the

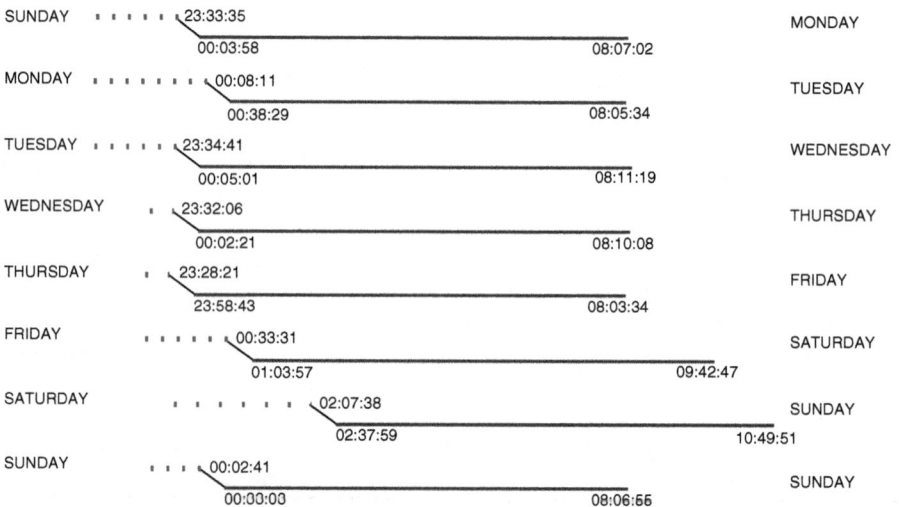

Fig. 5.1 User's behavior deducted from switch on/off time as portrayed by silent SMSs

20:09:39 ▪   ▪   ▪ ▪ ▪ ▪ ▪ ▪ 20:36:07

                    20:17:42 20:18:48   20:33:02 20:34:14

Fig. 5.2 Deviations from user's everyday route can be deducted from switch on/off time as portrayed by silent SMSs

time the victim switched the phone off. In our example, the first night the mobile phone is switched off sometime between 23.33 and 00.03 and was switched on back at 08.07 next morning, the other days of the week we have a similar time, while on weekend the pattern changes. In a different setup, as shown in Fig. 5.2, continuous SMSs of this type are sent. When the user gets in an area of no coverage (i.e. in an elevator, in a tunnel), we have delays in the delivery report. If the victim follows the same way to his/her work every day, crossing such a point, then the attacker is in place to know deviations of the victim from his/her everyday route. Finally, this technique can also be used by police forces before a raid in a suspect's house, or to force the suspect's mobile phone to transmit as to be easier identified among other phones when using an IMSI catcher as described in Chap. 3.

Leaving aside malicious actions regarding interception of messages, there is also the lawful interception where operators are required to help law enforcement authorities. They can disclose messages that have been sent and received by a particular mobile. Even without this cooperation, however, there are techniques that can retrieve messages already deleted from the memory of the mobile and SIM card, information which is of high value for any forensic investigation. Such processes depend on the particular phone model and are not always successful. More details will be given in the next chapter that deals with forensics.

5.4   Integrity Issues

SMS spoofing takes place when a message sender masquerades as another sender. It is used to avoid paying messages, hide the fact that a message is spam or fraudulent, and to avoid detection. The simplest way to send a (normally billed) message with a forged originating-sending number is via any mass messaging service (bulk SMS service). Using the Internet, the attacker can send a message picking up whatever id he wants (usually there is a limit for a string up to 11 characters or a number up to 16 digits). If the attacker chooses a phone number that is already present in the contacts catalog of a device, then when that message arrives, the victim's phone screen will match the number to the entry in the catalog showing that the message is arriving by a trusted person and not by a third malicious user.

Since this service was abused many times in the past, some bulk operators have banned the numerical-only id option. As such, users are forced to enter non-arithmetic characters too in the originator's id. Fraudsters soon found out a way to overcome this. They are using the letter "O" instead of the digit "0" or the letter "I" instead of the digit "1." In addition to that, as it was already explained in Chap. 4, attackers can always count on the catalog's functionality. Assume that number 6912345678 is present in your contacts list. The attacker can use the bulk sending service and send you a message picking up the originator's number AA345678. He has used non-arithmetic characters (AA) so he can escape the bulk operator's test and send the message. Your phone, however will match only the last digits and consequently it will show the name of the contact having the number 6912345678 instead of the number, as it was described in Chap. 4. You will see a familiar name you have in your contacts and you will proceed with opening and reading the message.

At a more advanced level, one can be directly connected to a messaging center (with a leased line, a simple dial-up call, or the Internet) and using the appropriate interface supported by the system (SMPP, EMI/UCP, TAP, etc.) send the message. In some SMSC protocols, the original sender of the message is identified in a specific field of the short message. Using spoofing techniques and giving a false number in that field may cause the message to appear as one coming from a different mobile phone [15, 16].

Usually, the messaging centers used by individual companies sending bulk messages are in a foreign country and possibly in the other side of the globe. Technically savvy users can thus check the number of the serving center to verify if it matches with the numbers of the sending party's operator. For example, if the originating number appears to be Greek, but the serving center's number appears to be from Finland, then this is a hint that something is wrong. Checking messaging center's number can be done in many ways. The first way is to use a handset that indicates not only the number of the sender but also the message center number (this was possible in many older phones). Another way is to connect the mobile to the PC and examine the source of the message. Finally, the user can use programs and applications specifically written for this task for his mobile. Even easier, a highly revealing factor is the time stamp of the message. Indeed, messages are timestamped with the time of the country where the SMS switching center of the home operator resides rather than local time of the recipient. So, if a message arrives with, say 2 h difference from the local time, from a number appearing to be in the same country (that should hence have the same time), the user should again be alarmed.

Moreover, some SMS can affect the integrity of the data in the mobile phone itself. They can, for example, activate or deactivate indicators presenting information about the status of the connection (such as the indication of new voice messages), or even cause a situation where these indicators are permanently lit without the user being able to switch them off. We examined such an example in the previous chapter.

Similarly, a very challenging and quite advanced perspective is the ability to send special binary messages. These messages do not appear to the user, but are directed to the SIM card or the phone itself directly, without alerting the user. One immediate advantage of this application is the ability of remote (OTA—over the air) modification, customization, and upgrading of the mobile phone and its software–firmware without the need for the subscriber to visit the operator's stores. On the other hand, if the existing security mechanisms are overcome (or even worse, if the attack is made with "internal" help), then the mobile goes to the full control of the attacker and allows both the downloading of malicious code and the interception of data, using SIM toolkit functionality. Finally, let's not forget the multimedia enhanced SMSs, too, known as MMS. Spoofed MMS can lead the user to download and possibly install malicious software in his/her phone.

5.5   Other Security Issues

In this section, we will mention some further features of short message services. Although they are part of the standards and there is almost always a business case for their presence and proper user, a malicious user can manipulate them to his advantage. Exploiting various techniques, he can indirectly affect the principle of integrity.

The first feature is the ability to send messages that can be automatically deleted or replaced later by another message, which will appear in the same memory location. The obvious reasons for such a category of messages were intended for use in information services (e.g. news, weather forecast, stock exchange, etc.). It would indeed be very convenient if there was such a text message permanently stored in the same position. Reading them, the user would receive the latest information every day, without having each time to delete the message and wait for the arrival of the next one. It is evident that there exists a clear potential for malicious use of this service by someone who wishes to change the content of a message already sent. In a simple example, let's assume that I have sent you an SMS for a meeting at 13.00. It happens that I am late, so I am changing the same SMS to write that the meeting is at 14.00. When I arrive, 1 hour late and being asked why I was late, I am prompting back to the SMS: "I never said that the meeting was at 13.00, check your SMS again!". And indeed, the SMS magically states 14.00 as the time of the meeting! The actual technical way to do it was described in Sect. 4.6.6.

The second characteristic that can be used in a malicious way is that of sending messages that present themselves directly on the mobile screen (flash SMS or technically more accurately, class 0 messages), without the recipient having to press any keys and perform the standard function of reading the message. They frequently appear in place of the logo of the network, or below it or in a window opened for this purpose. They remain visible until the user presses a key and then they are discarded. In a way, these messages arrive already "opened."

This "directness" and their differentiation from standard SMSs can trick the unsuspecting user, making him believe that these messages are actually sent from the provider. Using social engineering techniques, the attacker is able to persuade the victim to perform a function or call a number on the pretext that it is a suggestion from the provider (Fig. 5.3). Imagine receiving such a message, urging you to call a specific number, for maintenance reasons, and being labeled as coming from the provider. Many users would fall for that, calling a premium rate services number, with the fraudster receiving commission earnings.

Fig. 5.3 Flash message
masquerading as to be
appearing by the provider

Apart from the threats to the integrity of their content, short text messages can be used to affect the integrity of other mechanisms. A typical example in this case would be the mass sending of SMSs to influence the outcome of a contest or televoting conducted using SMS. In 2010, there were such allegations regarding the selection of a singer to represent her country in that year's Eurovision song contest [17]. There were accusations from both sides that there was some kind of tampering with the SMS voting process.

If this appears to be of little importance, the user should bear in mind that more serious attacks could target mobile banking where SMSs are being used by the bank to inform, or alert users. Many online banking systems use SMS as an "out of channel" authentication means, offering an extra layer of security. With spoofed SMS, this security measure can be less effective.

5.6   SMS Spam

In this section, we will focus on SMS spam. According to the latest forecasts, SMS will remain a significant source of revenues and traffic for mobile operators on a global basis until at least 2015. Global SMS revenues are forecast to rise to US$136.9 billion in 2015 from US$105.5 billion in 2010, while global SMS traffic is expected to increase to 8.7 trillion messages in 2015 from 5 trillion messages in 2010 [18].

The total messaging market in USA will rise from 2.3 trillion messages in 2010 to 3.5 trillion messages in 2015 [19]. An ever-growing percentage of messages are spam. Spam could be defined as unsolicited commercial messages from unknown originators. Marketing created spam as a method to reach out to a wider audience than what was traditionally considered. In a digital world, spam started as a plain text email, created to push the recipient to a website to hopefully make a transaction, buying a good or service.

There is no international agreed definition of what is and what constitutes illegal spam. There are different definitions provided by Australia, European Union, and USA. Australia: "Unsolicited commercial electronic messages", European Union: "Electronic mail for the purposes of direct marketing," and USA: "Commercial Electronic Mail Message" [20].

As the growth of subscriber and messaging volumes continue, unwanted and malicious messages become more frequent. An estimated 72 % of all mobile phone subscribers worldwide are active users of SMS. This presents a major problem for global messaging, as all users are a target for spam, spoofing, and other SMS-related scams. The impact of SMS spam can degrade network performance and quality of service (QoS), resulting in the loss of revenue and increased customers' concerns. As technology evolved, richer media was introduced into the mix (such as MMS — multimedia messaging service). These rich-media brought in Trojans (malicious software that is masquerading under a legitimate looking one) and opened back doors for damaging devices [21].

For Europe specifically, according to an investigation by UK Security-Park in 2009, European operators admit that up to 20 % of their users are affected by SMS Spam yet 83 % do not have a filtering system in place to prevent its effects on customers. And the result of a survey by Coleman Parkes in 2008 was that while the top 12 mobile operators across Europe anticipate even more fraudulent SMS driven by the mass adoption of mobile social networking and mobile email, only two of these operators are actually looking into ways to combat the problem. A recent survey done by M:Metrics and Airwide Solutions revealed that on average 12 % of people in the five largest European countries are receiving SMS messages from companies without permission (SMS Spam), a rate that grew to 21.3 % from June 07 to June 08 [21].

Spam has become a serious problem because it is profitable and can be almost completely anonymous. This problem needs a multi-faceted approach. Besides behavioral, technical, and economic measures, anti-spam legislation is meant to work complementarily to them. Today's worldwide legislative coverage of spam is heterogeneous (legislation in different countries is presented in [15]). While some countries have not introduced any anti-spam legislation at all, others have arrived at some degree of legislation [22]. In the USA, recipients of SMS spam can file a complaint with the FCC (Federal Communications Commission).

One of the biggest sources of SMS spam is number harvesting carried out by Internet sites offering "free" ring tone downloads. In order to proceed to the download, users must provide their phone's number, which in turn is used to send frequent advertising messages to the phone. Once again, lack of user awareness is exploited.

Even though there are lots of email anti-spam solutions, when it comes to SMS spam confrontation, there are some issues. For example, SMSs are usually shorter than email messages. Only 160 characters are allowed in a standard SMS text, and that could be a problem because using fewer words means less information to work with. On top of this, using acronyms and concatenated words makes it more difficult to create rules [23, 24]. Content-based spam filters can be built manually, by pointing the set of attributes that define spam messages. These are often called heuristic filters, and some popular filters like [25, 26] have been based on this idea for years. Content-based filters can also be built by using Machine Learning techniques applied to a set of preclassified messages [27]. These so-called Bayesian filters are very accurate according to recent statistics, and their applicability to SMS spam seems immediate. This is why several research tests are conducted using different attributes' definitions, learning algorithms, and a suitable evaluation method [22, 28, 29] as well as hybrid methods [30]. Many scientists believe that email filtering techniques require some adaptation to reach good levels of performance on SMS spam, especially regarding message representation [31]. They propose using three stages regarding the mapping between email and SMS anti-spam techniques: Data collection & Processing, Spam filters, and Evaluation.

To protect the SMS gateway from attacks, the messages or connections destined to the gateway should be strongly authenticated. This way the only senders of the messages would be the application server and the SMSC. There would not be any chance of spoofing and thus sending masses of bulk messages to the gateway from a fake SMSC. In the gateway market, there are several gateway products, which offer the ability to cryptographically protect the connection between the application server and the gateway [16] and to filter traffic. One of the drawbacks of existing solutions, however, is that they often look for topical terms (such as pharmaceutical product names) or phrases to identify spam messages. On the other hand, some of the legitimate SMS messages that contain such blacklisted words can be mistakenly classified as spam. This could happen more frequently with SMS messages than with emails due to their smaller size and simpler content. Many anti-spam solutions have been suggested based on a challenge-response protocol, while others suggest solutions which combine white/black listing and challenge-response methods.

In a process level, there are carriers that allow subscribers to report spam by forwarding the spam messages to them. Some spam countermeasures depend on detection, and there are two developments in that area: a GSMA pilot spam reporting program, and the development of Open Mobile Alliance (OMA) standards for mobile spam reporting. Another approach to reducing SMS spam that is offered by some carriers involves creating an alias address rather than using the cell phone's number as a text message address. Only messages sent to the alias are delivered; messages sent to the phone's number are discarded. Finally, most cell phone providers offer the option of completely disabling all text messaging services on a user's account. This extreme solution, however, is satisfactory only for those users who have neither the need nor the desire to utilize SMS at all.

A promising solution for the problem is found in [32]. The developed application runs on the mobile phone, using the service responsible for storing and forwarding the SMS messages, and is able to perform checks against common tactics and techniques employed by spammers.

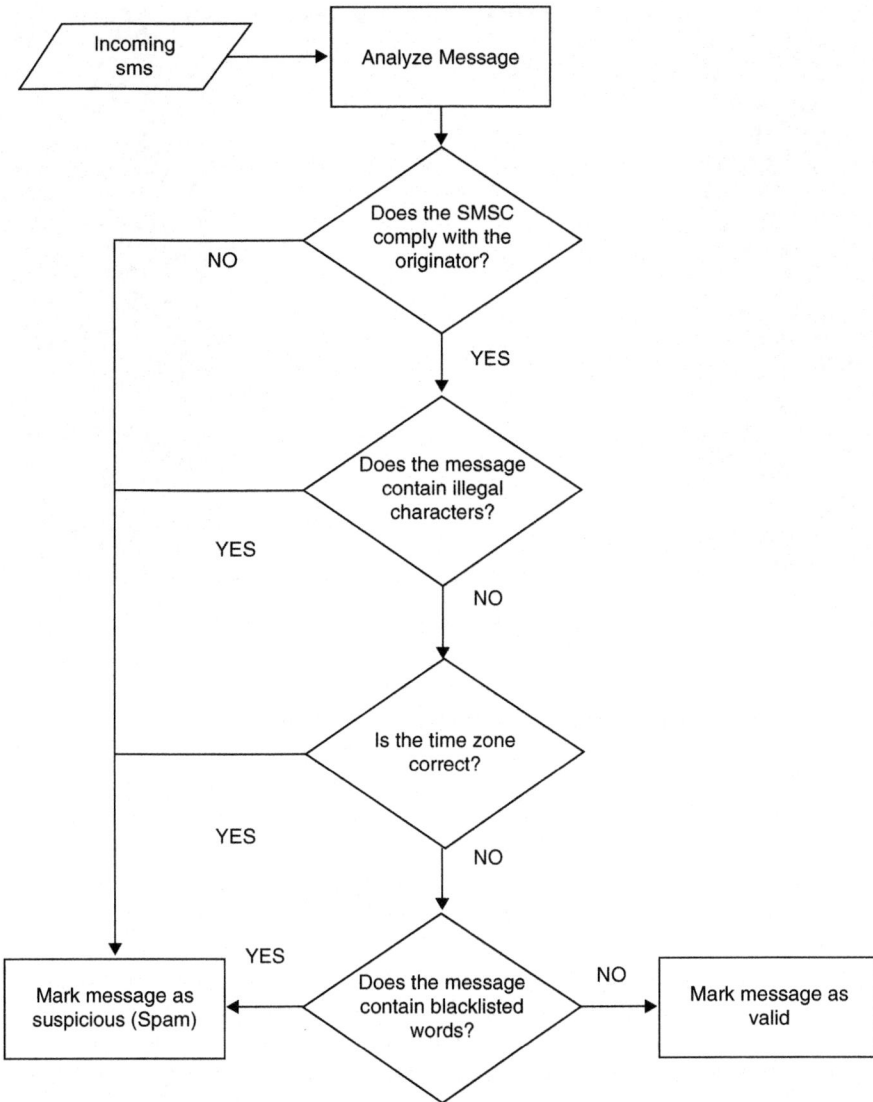

Fig. 5.4 The algorithm

Initially, the application monitors incoming SMS messages and records in a lightweight database the SMSC of every sender. It then applies the following rules (as seen in Fig. 5.4) to distinguish between legitimate and unsolicited communication:

1. Whenever a new SMS arrives, the originator number is checked to determine whether it has sent an SMS before. The developed application stores in its local database, among other pieces of information (Fig. 5.5), the SMSC number and the originator's number for each received SMS so as to enable the extraction of the international prefix.

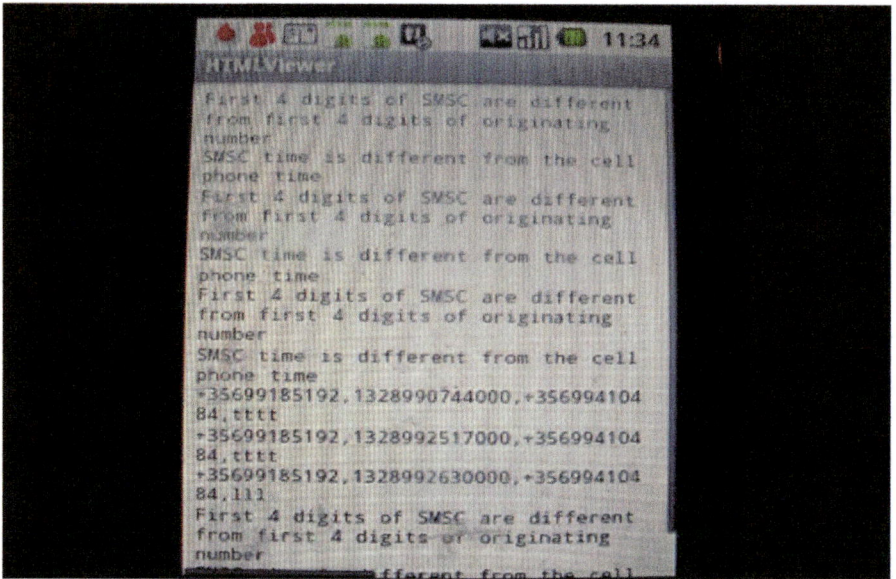

Fig. 5.5 Internal database of application where numbers are logged

If the number already exists in the database, the system compares the SMSC of the received message to that in the database record containing the same originator number. Should all comparisons return a perfect match, nothing is reported back to the user. Nevertheless, if differences exist (same originator sending an SMS from an SMSC different than the one used previously), the message is labeled as suspicious and the user is informed with the following message: Number (in full) +35961234567 has sent you x messages using SMSC + 359xxxxxxx and y messages using SMSC + 359yyyyyyy while now is using SMSC + 359zzzzzzz.

2. Usually, the first few digits of an SMSC match the respective digits of the sender's number. This leads to the provider's identification via the numbering plan, unless number portability is in effect. In this case, the matching digits are reduced to country prefix and possibly to mobile services pre_x (e.g. +3069 for Greece, where +30 is the international prefix and 69 is the national mobile phone services prefix according to the numbering plan). As such, a message sent from an SMSC belonging to another country than the sender's country is most probably spam, sent from some bulk SMS service situated abroad. Consequently, subscribers who reside in countries with SMSCs that routinely send spam messages can only be protected by black-listing all suspicious SMSCs.

3. The sender's identification is checked to determine if it is a purely numerical one, or it contains other characters as well. A message with a non-numerical sender ID has a great chance of being spam. As mentioned in Chap. 4, bulk SMS providers used to allow the sender to freely choose his sender ID, greatly facilitating spoofing attacks. Since this was greatly abused, some providers enforced the mandatory inclusion of at least one non-digit character in the sender ID field. Spammers soon started using letters that resemble numbers. For instance, the letter "O" is used instead of the number 0 or character "l" used instead of number 1.

Such characters combined with numerical digits in the sender's ID are a further indication of a possible spoof attempt. Therefore, only digits and the "+" sign are considered valid. Anything else will flag the message as being suspicious. Another relevant "feature" that can be manipulated by fraudsters and was again seen in Chap. 4 has to do with the way mobile phones map the numbers dialed (or the incoming numbers calling, or the ID of the sender of an SMS) to the respective entries in their catalog list. The catalog application in most cell phones matches only the 6-8 last digits of a number in order to show the respective entry name. Let's assume that a user has an entry "Iosif" with number "+306912345678" in his phone. A spoofer sending an SMS with a sender ID of "AAA345678" will manage to get his message delivered to the victim's phone with the display showing "Iosif." Because of this, although the message has arrived from a completely unknown number, the display shows a contact in the user's list, thus making a social engineering attack trivial.

4. The time zone of the SMSC and that of the mobile phone can also be used as possible indicators of a spam message. Our system designates a message as possibly being spam if the sender's number is in the same country as the recipient but the time of the SMSC is more than 1 min ahead. This is an indication that the message is coming from a country with a different time zone than the country of the sender's number (the difference in country is the fact that is of interest) (Fig. 5.6). Unfortunately, the check cannot be extended to messages stamped with a time earlier in the past, since this can be appointed to delays and not necessarily a country with a different time zone.

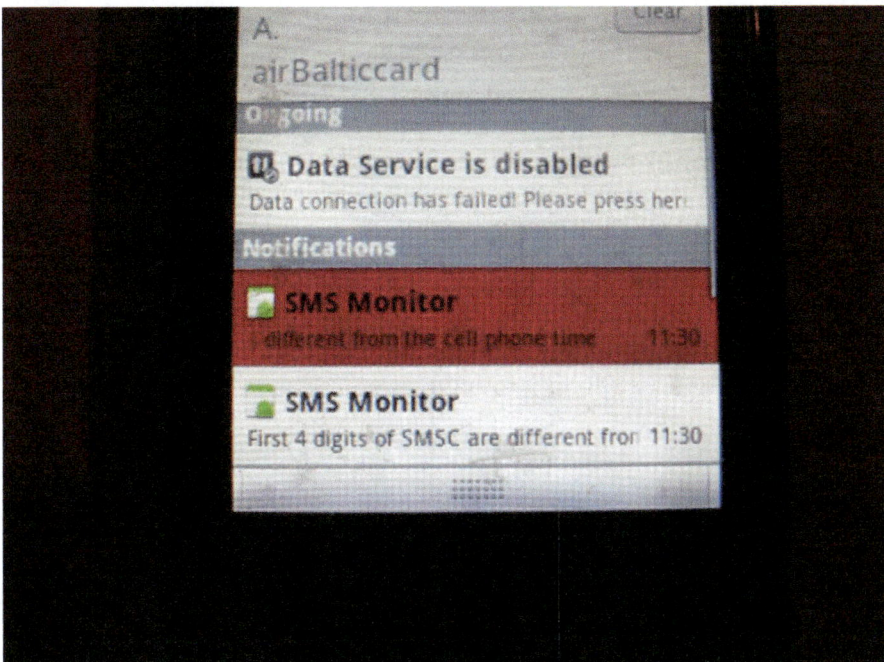

Fig. 5.6 Message marked as suspicious because of different time zone and different SMSC

5. Keywords blacklist. Blacklisted words such as "viagra," "replica," etc. immediately characterize a message as being questionable. New words can be added to the list by the user.
6. If applicable, it is checked whether the reply path field is being used and the message is classified accordingly. The specifications by the 3rd Generation Partnership Project (3GPP) allow a reply to an SMS to be sent from an SMSC other than that of the sender's. In practice, providers do not permit this functionality, however, it could potentially be exploited to perform an attack.
7. The application checks whether HTTP links exist in the SMS and informs the user accordingly, since such a link could point to malicious content.
8. In addition, the SMS protocol ID (TP-PID) [33] is checked. There are some special-purpose SMS messages, the so called "silent SMS" that get automatically deleted upon reception. These can be used for determining whether a mobile phone is switched on or not, without the user ever realizing this action taking place [13]. Despite the automatic deletion of the SMS, the application is able to notify the user that such an SMS was received. Furthermore, other techniques exploiting specific User Data Header (UDH) and User Data fields are also communicated to the user, since the application is checking for the presence of such information in the message.

Appropriate messages appear in the mobile phone's screen, if any of the above rules is satisfied, such as the existence of invalid characters, time inaccuracies, or blacklisted keywords, etc.).

5.7   Conclusion

Short text messages form the most widespread GSM service following basic service of voice communications. They are universally known and very popular among users because of their ease of use and immediacy. Unfortunately, they are facing a number of threats that affect their availability, confidentiality, and integrity. The advent of even more advanced capabilities and services, including use of electronic shopping (m-commerce) and banking transactions using mobile phones (m-banking), which largely rely on the ability to send and receive short text messages to authenticate the user, will raise even stronger security issues. It is therefore essential that the user is well aware and careful in the use of SMS until these issues are addressed to ensure the highest possible security level.

References

1. UK hails 10th birthday of SMS, December 2002. The Times of India. http://timesofindia. indiatimes.com/articleshow/30216466.cms
2. Americans sent 1 trillion SMS text messages in 2008. http://www.intomobile.com/2009/04/06/ americans-sent-1-trillion

3. 3GPP TS 03.40. Digital cellular telecommunications system (Phase 2+); Technical realization of the Short Message Service (SMS) Point-to-Point (PP)
4. TEKELEC (2007) SMS security: malicious attacks are just around the corner. Are you protected?
5. Androulidakis I (2009) Security in SMS. IT Security Professional Magazine 11:36–41
6. de Haas J (2001) Mobile security: SMS (& a little WAP), HAL200
7. Miller C, Mulliner C (2009) Fuzzing the phone in your phone. http://www.blackhat.com/presentations/bh-usa-09/MILLER/BHUSA09-Miller-FuzzingPhone-SLIDES.pdf
8. Mulliner C, Golde N, Seifert J-P (2011) SMS of death: from analyzing to attacking mobile phones on a large scale. In: 20th USENIX Security Symposium
9. Windows phone SMS attack discovered reboots device and disables messaging hub. http://www.winrumors.com/windows-phone-sms-attack-discovered-reboots-device-and-disables-messaging-hub, 2011
10. Engel T (2008) Remote SMS/MMS denial of service—"curse of silence" for Nokia S60 phones. http://berlin.ccc.de/~tobias/cursesms.txt
11. Enck W, Traynor P, McDaniel P, La Porta P. Exploiting open functionality in SMS capable cellular networks. In: 12th ACM Conference on Computer and Communications Security (CCS'05)
12. Agarwal N, Chandran-Wadia L, Apte V (2004) Capacity analysis of the GSM short message service, NCC2004
13. Androulidakis, C. Basios, A plain type of mobile attack: Compromise of user's privacy through a simple implementation method, Proceedings of 3rd International Conference on Communication Systems Software and Middleware (COMSWARE 2008), pp 465—470, 2008
14. Androulidakis I, Vlachos V, Chatzimisios P (2015) A methodology for testing battery deprivation for testing battery deprivation denial of service attacks in mobile phones. In: Information and Digital Technologies (IDT), 2015 Internal conference on, 7–9 July 2015, pp 6–10. doi:10.1109/DT.2015.7222942)
15. Network Security Solutions (2006) SMS vulnerabilities-XMS technology enabling mCommerce
16. Sillanpää A (2001) Mobile asset security and how to make money on it. In: T-110-501 Seminar on network security, pp 1–18
17. SMS-scandal overshadows Eurovision victory for Rivas. http://www.panarmenian.net/eng/culture/details/44736/, February 2010
18. Informa Telecoms & Media. SMS traffic growth driven by enterprise, emerging markets and social networks. 2011. http://www.informatm.com/itmgcontent/icoms/whats-new/20017843617.html, accessed on 28/04/2011
19. Global Information, Inc. (2011) Telecom & IT market report catalog
20. Bueti MC (2005) Anti-spam legislation. ITU. In: WSIS thematic meeting on cybersecurity
21. Airwide Solutions, Inc. (2009) Taking the challenge of mobile messaging abuse
22. Schryen G (2007) Anti-SPAM legislation: an analysis of laws and their effectiveness. ICT Law 16(1):17–32
23. Hidalgo JM, Cajigas Bringas G, Sanz EP, Garcia FC (2006) Content based SMS Spam filtering. In: Proceedings of the 2006 ACM symposium on document engineering, pp 1–8
24. Hidalgo JMG, Sanz EP, Cormack GV (2005) Spam filtering for short messages. In: Proceedings of the sixteenth ACM conference on conference on information and knowledge management, pp 1–8
25. The Apache SpamAssassin Project: SpamAssassin Guide, pp 1–5, accessed on 28/04/2011
26. WebGate (2009) SMS Spam manager guide, accessed on 28/04/2011
27. Sabri AT, Mohammads AH, Al-Shargabi B, Hamdeh MA (2010) Developing new continuous learning approach for spam detection using Artificial Neural Network (CLA_ANN). Eur J Sci Res 42(3):525–535, ISSN 1450-216X
28. Androutsopoulos I, Koutsias J, Chandrinos KV, Spyropoulos CD (2000) An experimental comparison of naive Bayesian and keyword-based anti-spam filtering with personal e-mail messages. In: SIGIR'00: Proceedings of the 23rd annual international ACM SIGIR conference on Research and development in information retrieval, ACM, New York, pp 160–167

29. Androutsopoulos I, Koutsias J, Chandrinos KV, Spyropoulos CD (2000) An evaluation of naïve Bayesian Anti-Spam filtering. In: Potamias G, Moustakis V, van Someren M (eds) Proceedings of the workshop on machine learning in the new information age, 11th European Conference on Machine Learning, Barcelona, Spain, pp 9–17
30. Yoon JW, Kim H, Huh JH (2010) Hybrid spam filtering for mobile communication. Computers & Security 29(4):446–459
31. Cormack GV, Gómez Hidalgo JM, Puertas Sánz E (2007) Feature engineering for mobile (SMS) spam filtering. In: SIGIR'07, pp 1–2
32. Androulidakis I, Vlachos V, Papanikolaou A (2013) FIMESS: filtering mobile external SMS spam. In: Proceedings of the 6th Balkan Conference in Informatics (BCI '13). ACM, New York, pp 221-227. doi:http://dx.doi.org/10.1145/2490257.2490288
33. TS 23.040, 3rd Generation Partnership Project (3GPP). Technical realization of the short message service (SMS) September 2010. Release 9

Chapter 6
Mobile Phone Forensics

Abstract More than often, in our days, cases go to trial and crimes are solved with the help of evidence taken from the mobile phones and their use. Typical examples of such evidence are the mobile phone's location (based on the serving base station and/or GPS data), the contacts correlation (based on incoming and outgoing calls), communication content (based on messages and emails send/received/stored), and so on. In this chapter, we will analyze the subject of digital evidence from mobile phones from both the theoretical and the technical side.

Keywords Mobile phone forensics • Mobile phone crime • Mobile phone evidence • IOCE • ACPO • Data preservation • SIM Card • GSM 11.11 • Files In SIM • JTAG

6.1 Introduction

"The mobile phone betrayed the criminals… ",
"73 mobile phones and SIM cards were found in their hideout…",
"The fatal error with the stolen mobile phone…"

The right use of evidence forms an important judicial process that significantly helps a case's hearing. The digital forensics analyst assembles evidence from the crime scene, evaluates its importance, and analyzes and presents the data in the court. In analogy to "classical" forensics, digital evidence analysis takes place using data extracted from any kind of digital electronic device [1]. Consequently, in case a person is involved in an illegal activity and has used such a device, it is more than likely to have left digital traces which constitute valuable material for the law authorities [2].

One of the most characteristic digital devices nowadays is the mobile phone. The mobile phone's penetration in daily life and its universal use are a given fact. Modern devices with advanced functionality rapidly converge to computers as discussed in previous chapters. As such, they are used by their owners not only for communication but also for data storage, organization, and processing as well as for internet browsing. As time goes by, more and more data are collected from their use, strongly correlated to their owner. Such data can be found in the mobile phones themselves as well as in the computer systems of the telecom providers' networks

© Springer International Publishing Switzerland 2016
I.I. Androulidakis, *Mobile Phone Security and Forensics*,
DOI 10.1007/978-3-319-29742-2_6

and infrastructure. Attempting a farfetched but interesting approach, we can assume that mobile phones constitute a "biometric" characteristic since their use is absolutely individualized for each user.

6.2   Crime and Mobile Phones

GSM/UMTS networks serve more than four billion users worldwide. Obviously, criminals are also included among them. They are using the technology for personal gain and "involve" mobile phones in various ways in illegal and criminal activities. Before banning anonymous service, prepaid cards and handsets were a common communication method for any kind of criminal activity, ranging from drug dealing to distraction (e.g. calling a guard in order to distract his attention). Moreover, being particularly small devices with an increased monetary value, mobile phones are an easy and frequent target for thieves. Finally, another daily phenomenon is harassment and bullying call cases of every kind as well as threatening calls.

In more serious cases, we must also consider the physical security factor. This factor can easily slip our attention if we consider that digital evidence collection is always a risk-free procedure, since it involves only electronic devices. A mobile phone can transform into a detonating mechanism with triggering capability from every point of the world, using a simple incoming call (Fig. 6.1). Respectively, the bombs in Madrid's metro in 2004 were triggered using a mobile phone's alarm clock. Moreover, apart from mobile phones-bombs, there have appeared mobile phones—weapons and mobile phones-tasers.

At the same time, focusing in "white collar" crime level and with the mobile phones in presence, we observe all kinds of telecommunications fraud, personal data theft, identity infringement-theft, industrial espionage (using the memory of the mobile phone or the phone itself as a bugging device to intercept and to transfer

Fig. 6.1 Mobile Phone Detonator. We have a clear view of the connection with the mobile phone using the input/output port

data and commercial secrets), and so on. The future mobile commerce options, using mobile phones to buy goods and services, will obviously make the problem more immense, providing more space for criminal activities.

6.3   The Evidence

As stated, mobile phones provide a continuous flow of data and information concerning their user and his behavior. Where all this information resides to? Not only in the mobile phone itself but also in the computer systems and networks of the providers. Specifically for the device itself, evidence can be found in its internal memory, the SIM card, and the possible external memory cards [3]. We shouldn't forget the case where mobile phones are synchronized or connected in any way with a computer; therefore, evidence of mobile phone usage can be found in the user's computer too.

Typical data to look for than can prove suitable evidence are the following

- Phone catalog and contacts
- Incoming/outgoing/missed calls
- Incoming and outgoing text messages and MMS
- Sound recordings/Vocal notes/Calling sounds
- Photographs, Video, Graphics, Workspace
- Calendar, Alarm clocks/Reminders, To-do lists
- Written texts/memos
- Emails stored in the phone
- Websites visited using the mobile phone
- Documents and files of any type
- User identifiers (e.g. PIN)
- Device identifiers (e.g. IMEI)
- Catalog of networks used
- Geographic information (e.g. from embedded GPS receivers)
- Words added to the predictive text system's database
- GPRS, WAP, and Internet settings

A particularly important differentiating factor concerning computer forensics is the fact that mobile phones were traditionally more "closed" systems, allowing less user control to their core functionality and their operating system. It was difficult to precisely know what data was stored where. Even with modern open operating systems used in smartphones, there is considerably less information and experience so far regarding their internal workings. Thus, application and data traces remain in memory, possibly in areas where the user cannot have access to erase them. As an example, simply deleting a piece of information (e.g. an SMS) does not always mean that the information was permanently erased. Rather, the specific memory area is marked as free for further storage. If the information itself is not overwritten, it continues to exist in memory, even though the user believes that he has erased it.

6.4   Fundamental Questions and Problems

As it is the case with traditional evidence and forensics, the specialized profession-als should follow strict and specific techniques for data extraction, maintenance, analysis, and presentation of digital evidence. Later in the chapter, we will cover the process and the specialized software and hardware tools used for the analysis. Before that, we will present some fundamental questions arising during the process of digital forensics in mobile phones [4, 5].

The main question to answer is whether the mobile telephone was used for crimi-nal/illegal activity. In case the answer is positive, can we extract relevant data to be used as evidence?

Certain problems manifest in the whole process: Was the specific suspect the phone user? In other words, the device was in the crime scene … but who was the user at that moment? Was it its owner or a third person? What happens if the phone has been borrowed/given away/sold to another person? Even worse, what happens if the culprit deliberately abandons the device to be found (and "confiscated") from some unsuspecting victim? This could turn authorities' attention toward the wrong person, misleading the whole search. In addition, how is it proven that the incrimi-nating evidence stems from mobile phone owner's use and not from some other malicious activity? For example, a data transfer can have taken place unbeknown to the user with some kind of intrusion. Technical ones include Bluetooth, malicious software, virus, IMEI cloning, or even (in older technologies) SIM card cloning, and so on. In the nontechnical ones, having lent the mobile phone or having lost it, even for a few minutes only, can lead to data planting.

It is also possible that records in telecom operator's systems are not complete, or that have not been logged at all (because of system overload, malfunction and so on). Maybe the data retention period for preservation of data has elapsed resulting in evidence being lost. Specifically for geographic data based on base station cover-age and not GPS, a phone can be registered and served from a cell far away from its actual position, should the local cell be overloaded/damaged. Thus the mobile phone user can wrongly be considered to be in a specific crime area. Finally what is the possibility of a simple human error (e.g. a wrong call at the wrong time to the wrong person, incriminating a completely innocent person)? Even though it may resemble a film script, statistically everything is possible.

Apart from the digital space, the examination should take into consideration the preservation requirements of other (not digital) evidence which can coexist (DNA, imprints, drugs, guns, etc.). Therefore, the examination sequence is important (e.g. touching a device with naked hands can destroy fingerprints and reversely, in the process of extracting fingerprints from a device, it is possible to accidently delete digital evidence that potentially existed).

The reader can now understand how crucial the evidence analysis procedure is and how careful the analyst must be. She has to examine every possible parameter that led to the presence of that specific evidence in the device. Simply extracting data is not enough. Only a multilayered and thorough correlation combining data from various sources (e.g. payphone logs, prepaid card logs, credit card logs etc.) leads to usable evidence.

6.5   Forensic Procedures

6.5.1   Introduction

In this section, we will be focusing on the nontechnical details of mobile phones evidence collection and analysis. We will cover procedures for evidence Collection, Preservation, Examination, and Analysis and finally for the writing of the Report. We will rely on the best practices that IOCE (International Organization on Computer Evidence Guidelines for Best Practice in the Forensic Examination of Digital Technology) proposes for digital evidence. They state the following [6]:

- When dealing with digital evidence, all of the general forensic and procedural principles must be applied.
- Upon seizing digital evidence, actions taken should not change that evidence.
- When it is necessary for a person to access original digital evidence, that person should be trained for the purpose.
- All activity relating to the seizure, access, storage, or transfer of digital evidence must be fully documented, preserved, and available for review.
- An individual is responsible for all actions taken with respect to digital evidence while the digital evidence is in their possession.
- Any agency, which is responsible for seizing, accessing, storing or transferring digital evidence is responsible for compliance with these principles.

We will also use ACPO's (Association of Chief Police Officers) guidelines which in turn has also published its own best practices on the object (Good Practice Guide for Computer based Electronic Evidence) focusing in four principles [7]:

1. No action taken by law enforcement agencies or their agents should change data held on a computer or storage media which may subsequently be relied upon in court.
2. In exceptional circumstances, where a person finds it necessary to access original data held on a computer or on storage media, that person must be competent to do so and be able to give evidence explaining the relevance and the implications of their actions.
3. An audit trail or other record of all processes applied to computer-based electronic evidence should be created and preserved. An independent third party should be able to examine those processes and achieve the same result.
4. The person in charge of the investigation (the case officer) has overall responsibility for ensuring that the law and these principles are adhered to.

A more technical introduction is given in [8, 9].

6.5.2   In General

During mobile phone's forensics examination, the evidence should always be extracted maintaining the data integrity. General principles of forensics that concern classic evidence are also in effect for digital evidence. Reliable procedures for securing the scene, documenting/photographing it, evidence collection and proper packing, transfer, and storage should be followed. Each action relative to the data confiscation, access, transfer, and storage should be completely logged and available for possible audit. During the confiscation, the area and evidence are photographed and/or videotaped. It is very important to photograph the mobile phone's screen contents as it was found. After confiscation, the evidence should not be altered. Consequently, the analyst should have proper training while the procedures she follows should be possible to be repeated producing the same results in order to prove their validity. At the same time, in all stages, each evidence holder is responsible for anything that happens to it. Finally, the procedure should be completed as quickly as possible because if the mobile phone remains for a long period in the forensics lab, there is the danger of evidence alteration accusations from the defended. More details can be found in [10, 11]

6.5.3   Training and Competence

All involved persons in the process should be suitably trained in order to perform and document their actions. Mobile phone technology evolves so fast that using the best equipment alone does not guarantee the success if it is not accompanied by regular and thorough education of the examiner.

There are different kinds of responsibilities and training, according to the involved personnel. Police officers who act on the scene of the crime have different responsibilities than analysts. The first ones should secure the scene, log and collect the physical evidence (mobile phones devices, cables, chargers etc.), and pack and safely transfer them to the lab. They shouldn't do the analysis themselves since they do not have the necessary know-how and the specialized equipment. The analysts, who are the ones responsible for the analysis, perform the data extraction and the examination of it, finally writing the corresponding report. They are obligated to remain informed and to follow the technology pace and the scientific research evolution. It is also very important to practice with a same mobile phone and the same software before performing the analysis in the actual confiscated mobile phone. In any case, their work doesn't stop in the report but they must also safeguard the evidence for possible future need and access.

The complete responsibility for the suitable procedures observation lies with the Head Officer. He must ensure that all the involved personnel have the proper training in order to complete the corresponding procedures. Following, during the analysis, he must maintain continuous communication with the analyst and provide guidance since the Head is the only one who knows the whole case. If this communication is disturbed, it is possible to lose data which can have decisive importance for the trial. On the other hand, the expert in charge of the forensics research decides for the suitable methodology she will follow depending on the case's severity. For a simple case, she can use simple software data recovery from the mobile phone and the SIM card. For more difficult cases, it is possible to use memory dump and search for evidence using special hardware and software. For cases of "life or death" situations and when all the other methods fail, she can physically remove the memory and try reading the chip's contents. These processes will be technically reviewed in the next sections.

6.5.4 The Analysis Procedure Itself

During the procedure of data extraction, different situations are possible. Mobile phone access might be impossible (therefore the only data that can be used are based on the provider's help), can be temporary, or in the best case scenario the mobile phone has already been confiscated and is in the examiner's lab. In the rest of the chapter, we will assume that the mobile phone has already been confiscated.

In the lab, during the data extraction process the device is initially recognized and the selection of tools and methodologies that will be used follows. The expert seeks information for "who", "where", "when" and "why" by examining data stemming not only from the mobile phone's ownership but also from its possession (it is possible that the mobile phone has a different owner than the holder of the phone at the moment of the confiscation). This is performed by examining data, applications and files, categorizing the time line of events, without overlooking hidden evidence that possibly exists.

The research plan is carried out in such a way as to avoid data loss as much as possible. In general, if the mobile phone is already switched on, the memory's data are examined first and then the SIM card's data (before both, mobile phone's isolation from the network should be ensured as we will see below). This way, the examination can proceed immediately without depending on the holder's or provider's collaboration. This point plays an important role. If the mobile phone is switched off and the holder does not reveal the PIN code, the authorities have to ask for the provider's help. If the SIM card is issued by a foreign country, then the communication with the proper authorities and the corresponding operator can delay the process for months.

If the mobile phone was deactivated during the confiscation, then SIM analysis will take place first. Activating the phone, even with the same SIM, can cause data alteration (e.g. in the Location Area Information file). Even worse, placing a different SIM in certain cases can lead to data deletion (e.g. the last calls history).

As a general principle, it is good to photograph or videotape each step of the process, time stamping it. The examiner's actions are recorded in the corresponding log. The log includes details for the time, the action, and its result. A third person following the same instructions should reach to the same result. Even for automated data analysis, videotaping is essential in order to prove that the correct way of operation was used. It can also be used to verify the results. The tools and the software version of software used should also be documented and included in the final report.

6.5.5 Data Preservation and Isolation from the Network

The main principle of forensics is the preservation of data indicating that data should not be altered at all to allow its use in Court. For this purpose, it is advised to solely use special software as nonqualified and unchecked software can potentially write/alter data on the phone destroying the integrity of the evidence. Unfortunately, the phone may often not be compatible with specialized software but only with generic software. In such cases, operating methods should be checked on a same model test telephone. In addition, it is attempted to use this non specialized software as late as possible in the process.

A case can literally be "lost" due to mistakes in receiving and transferring data that destroy evidence. It is particularly important to avoid "contamination" from new phone calls, messages, and generally communication with the network. In addition to new data that may arrive (e.g., calls or messages), destruction mechanisms can be enabled including deletion of data or locking of the device (e.g. with a specific incoming SMS and appropriate software running on the phone). For this reason, the device should be isolated from the network. If the device is in operation, then it is advisable to remain in this state and not be turned off by the analyst because as described earlier PIN might be needed later which will delay the process.

The best solution is to place the device in a special cage that isolates electromagnetic radiation (Faraday cage). This way, the device can be transported safely to the laboratory. For convenience, special smaller isolation bags are available but they cannot prevent signal penetration if the mobile phone gets near to a base station where the signal is very strong.

During the period in which the device is isolated from the network, the battery will run out very soon because the mobile searches for a network to connect to, constantly scanning the frequencies. It should therefore be charged regularly to avoid data loss. In this case, special care is needed to avoid RF signals entering via the charging cable. This is why a mobile power source, placed inside the Faraday cage, is the best solution. Once at the laboratory, analysis can take place in a special, shielded from electromagnetic radiation room. Since the cost of such a room is very high, there are Faraday cages, larger than the transportation ones, with special feeds for the hands. This way the examiner can manipulate the phone while keeping it still isolated from the network.

A more "aggressive" way is to use interference devices/jammers. The use of such devices may be illegal and besides they will inevitably cause problems to other nearby mobile phones. Specifically for modern mobile phones, another alternative is the activation of the airplane/flight mode. This function isolates only the transceiver of the phone, while other functions continue to run normally.

One more option, not necessarily being a good solution, is to use a specific SIM (correctly configured) that activates the mobile but does not allow network access. This solution enables transportation but there is always the possibility that changing the SIM will cause data alteration on the mobile, so it should be tested on a same mobile phone model beforehand. Finally, in collaboration with the operator, it is also possible to administratively isolate the cell from the network.

6.5.6 *Identification of the Phone*

Once the mobile phone has safely arrived in the forensics lab, the analyst's first task is to identify the exact model and type in order to assemble the necessary connection cables, chargers, manuals, and so on. An optical recognition of the mobile phone is often enough. It is likely, however, that the model is unknown or does not exist in the local market. There exist various mobile phone databases in the Internet with mobile phones' photographs; it is therefore possible to have a first identification effort. If even that fails, the process is differentiated depending on whether the mobile phone is activated or not. If not, then usually under the battery, its IMEI serial number is written along with other identification data such as the MAC address (Fig. 6.2). This allows the exact identification. If the mobile is switched on, then by typing *#06# as we have seen in Chap. 4, the IMEI is revealed. This method is the last choice because we want to have the minimal possible interaction with the mobile phone.

Having identified the phone, the examiner searches for its user manuals and its technical characteristics. Their careful reading is mandatory in order to realize whether the risk of data deletion (e.g. if SIM change takes place) is present. In any case, a check using a same test model is suggested. This way the examiner is assured for the proper operation of the tools she is going to use. This step is of course essential in case the examiner encounters for the first time a specific device. Having read the manual and being familiarized with the device's use, the examiner is in place to proceed to the next steps. She will define the storage area and extraction capabilities and will locate the available control and data interfaces. Adhering to the security and data integrity requirements, the examiner has to understand the device design and explore both electronic and mechanical parts and interfaces. At the same time, she will calculate the time needed and the cost of the procedure before finally extracting the data.

The connection with the mobile phone usually takes place with the respective cables while there are different protocols for the actual communication with the computer. Manufacturer's cables and not anonymous ones are the preferred way to connect the phone. Interconnection via Bluetooth is the most risky method and will

Fig. 6.2 IMEI and MAC address sticker under the battery

MAC:4C5499EF███

IMEI:3516020410███

S/N:NUA7NB10920███

CE0197 ①

FCC ID:QISU8120

MADE IN CHINA

eventually write data in the memory of the phone (key pairing). It must therefore be avoided. The same applies for Wi-Fi connection, while infrared has ceased to exist for quite some time.

6.5.7 Examination of the SIM Card and the Memory of the Phone

The device itself may be activated, deactivated, or even damaged. In addition. the phone itself and/or the SIM card may be protected by a password. In this case, as well if the suspect does not cooperate, the SIM's PIN should be sought from the provider. At this point, we must mention that initially the owner of the mobile phone is asked to reveal the PIN and other codes. He is never allowed to touch the mobile phone and type the code since he could try to delete data and incriminating evidence. A written statement may also be obtained on whether the cell is damaged or not as to avoid the accusation that it was destroyed during analysis.

Regarding the analysis of the SIM, there is a key differentiation from the analysis of evidence on computer hard drives. As we will see, the SIM provides access to the data stored in its memory through its own microcontroller. It is therefore not possible to "clone" it bit by bit. This means that the analysis takes place on the original exhibit rather than a copy. The ideal would be to have a way to get all the data on the spot and to calculate the corresponding hash. The hash is a unique fingerprint/signature for every file, achieved by special algorithms. Checking the hash values of two different files or data sets, we can easily certify that they are the same and that there are no differences at all, not even in a single bit. Unfortunately, it is completely impossible to "clone" the SIM memory. On the same wavelength, reading of certain elements inevitably causes their alteration (for example, when opening an unread message, its status will change to "read"). An easy method to combat against this fact is to film the process. In any case, contrary to the analysis of evidence on computers, in mobile phone forensics, there is generally more interaction. Methods damaging the physical integrity of the device (e.g. removal/unsoldering of circuits for external memory reading) can also be used.

The analysis of the contents of the SIM card includes technical details that will be examined in the next section. Of great importance is the counter status of these codes. If, for example the maximum number of wrong PUK entries is close, a further wrong PIN entry would lead to the card being self-invalidated and impossible to read. In addition, extracting the SIM card to examine it with a special reader requires almost always removing the battery of the mobile. Depending on the model, this may result in deletion of items such as time and date, either directly or indirectly after a short period. It is advisable, again, to have made a relative test with an identical device. The battery should be placed back again as soon as possible. If the deletion of data by removing the battery is inevitable, then the examiner should first complete the data analysis of the data in the mobile before removing the battery. Complete discharging of the battery may lead to the same problem; hence, special care for temporary charging should be taken.

Following the extraction of evidence with the use of automated tools, manual effort can verify the results and sometimes further discover evidence that for various reasons (e.g. insufficient device support) the software failed to find. Consulting the manual, the examiner can proceed with manual reading of the other elements as well as verification of the evidence already extracted, using the respective options of the phone. Photographing or filming the steps is highly recommended again. For this purpose, special holders exist that can facilitate the filming. Needless to say, that in cases where no compatible software for the analysis can be found, the manual method may be the only option available.

In conclusion, the evidence extracted must be verified. If possible, a cross check could be sought with the network operator. For example, calls on the mobile can be verified testing them against the calling logs of the network provider. Using scientifically proven procedures and thorough analysis, the examiner increases the importance and weight of the evidence since it can no longer be disputed.

6.5.8 Findings Report

The process ends with the Report of Findings which records both the process followed and the findings. In this report, initial data for the Agency, the case, the staff who dealt with it and the Head, as well as the dates of the events should appear. The presentation of the software and tools used and the methodology followed comes next. The planning of the methodology and the steps taken must be clearly formulated. The accompanying materials and equipment used are recorded in the relevant section. They are presented after the findings from both the automated extraction and the manual investigation and the report closes with conclusions. A good report will assist in the proper trial so the analyst should possess advanced writing skills too. In fact, an excellent technical part of the forensics process may lose its value if not presented in a well intelligible way on the report.

6.6 The SIM Card

In the rest of the chapter, we will focus on the technical details of mobile phone forensics. One of the fundamental elements that compose GSM/UMTS networks is the SIM card (named USIM in 3G networks) which apart from the main role of user identity authentication provides basic functionality to the mobile phone. The SIM (Subscriber Identity Module) is a smart card with an embedded microprocessor and 16–256 Kbytes of nonvolatile (EEPROM) memory. The smart card's operation is analytically described in the series of standard ISO/IEC 7816 [12]. In general, the microprocessor provides access in memory data and is responsible for the security, while the mobile phone communicates with the SIM according to specific standards like GSM 11.11 [13] and its descendents [14]. These standards ensure a relative uniformity which helps considerably in evidence extraction.

The main drawback of this design, in terms of forensics, is the lack of direct memory access. It is not possible to bypass the microprocessor since it is the one that provides the access to the memory. This makes it impossible to dump the memory contents of the SIM card, the same way as it is possible when cloning (copying bit to bit) a hard disk drive. The examiner who wants to analyze data kept in the SIM gets access using microprocessor commands, with the help of a smart card reader or by using the mobile phone itself (if the latter allows it, possibly through the use of special AT commands, as seen in Chap. 4). In any case, forensics software can make the process easier by providing a suitable graphical user interface and automating a series of tasks.

Speaking of software, there are different options, covering different phones, mainly depending on the phone's operating system. Various protocols and techniques are used to communicate with the phone and exchange data. It is quite interesting to note that they can produce different results, as they are subject to many restrictions [15–20]. A major consideration is also the fact that they are operating on the "logical" level and not the physical level of the memory.

Apart from the core function of data extraction, the forensics examiner mostly welcomes features such as the automation of the report writing with ready templates, the option to search in the memory with many different ways, and the included libraries of video and sound as to be able to easily reproduce the files found. The ability to produce hashes of the data, preserving their validity is also very important, on a forensics point of view. This functionality is usually missing from non-forensics software. Ease of use, repeatability, and validation options are of course needed. The scientific community should have tested the product and the manufacturer should be providing ample support.

New tools offer more functionality but still there is space for work in the forensics software arena. There is no super-tool and the examiner in many cases has to work with a compilation of programs and versions in order to be able to extract as much data as possible. Software should be tested extensively since there is always chance of faulty operation, either due to bugs or due to the huge volume of standards and the possible deviations that manufacturers sometimes take. A very often situation arises where the forensics software fails to properly read non-Latin character sets and the report contains different data than the phone. The experienced examiner can spot these shortcomings and find the best solution for each case, combining most of the times some manual work.

Getting back to evidence, as we all know, SIM card access is protected using the corresponding PIN code (4–8 numerical digits). In secondary level, the PIN2 code also exists for the management of certain phone characteristics (e.g. network settings, restriction of possible calls only to specific numbers, and so on). At this point, we must note that the right term as it is reported in standards is CHV (Card Holder Verification) and not the term PIN (Personal Identification Number) which has prevailed. The PIN is originally set by the manufacturer or the provider but it can also be changed by the subscriber. This can be done using phone's settings or using network codes (**04*old PIN*new PIN*new PIN# for PIN1 and ** 042*old PIN*new PIN*new PIN# for PIN2) as it was described in Chap. 4.

As some readers might have found out, following three erroneous efforts of PIN entry, the card gets blocked and does not accept any more PINs. This happens in order to protect the card from brute force attacks where someone would try (with software help), all the 10,000 possible combinations (0000-9999) for a 4 digit PIN. Fortunately, it is possible to unlock the card with the use of PUK (PIN Unblocking Key or, more formally Unblock CHV). As a matter of fact, to unblock PIN1 we use PUK1 and for unblocking PIN2 we use PUK2. PUKs consist of 8 digits and are once again set from the manufacturer or the provider (they can be changed too with network code **05*old PUK*new PUK*new PUK# for PUK1 and **052*old PUK*new PUK*new PUK# for PUK2). If somebody is decided to destroy the card, after 10 unsuccessful PUK password tries, the card renders itself useless and the user has to visit the provider to issue a new SIM.

The series of codes and passwords in SIMs is completed by the ADM code which is known only to the provider for every card. The code provides complete access in the card contents as well as their addition/modification or their deletion, even remotely using SIM toolkit functionality as was described in Chap. 5. For further information on SIM cards, the interested reader can consult [21–35].

6.7 Files Present in SIM Card

6.7.1 In General

GSM 11.11 [13] standard defines among other things, the existence of a directory structure containing specific files in the SIM. The user can think of it as the analogy of directories and files in hard disks. Without going into further details, it is enough to remember at this point that inside these specific directories, there exist files/memory slots where various information elements are stored. This structure is described in Fig. 6.3 where MF (master file) is the root directory, DF (dedicated file) are the subdirectories, and finally EF (elementary file) are the actual files which contain data. Different read, write, modification, and deletion authorization levels exist for these files. Certain files can be read without the need to enter the PIN code, others require PIN authentication while in the most important ones only the provider has access by using the suitable ADM code as it was written earlier.

As we have already discussed, the SIM includes files containing mobile phone's capabilities, the card's serial number, the list of the providers and their names, the default network, the default languages, the contacts list, incoming and outgoing messages, settings for message sending, a list of the last outgoing calls, and so on. There are also files present that host the network's subscriber permanent and temporary identity (IMSI-TMSI), the subscriber's coarse location (LAI), control channels (BCCH), and the current encryption key (Kc). In total, some 100 files are present. A particularly interesting fact is that besides these standardized files, each provider is free to use his own files too. This way, issuing the right commands to perform a brute force reading of all the memory areas (through SIM's microprocessor) or following provider's feedback more interesting data can be found in these non documented files.

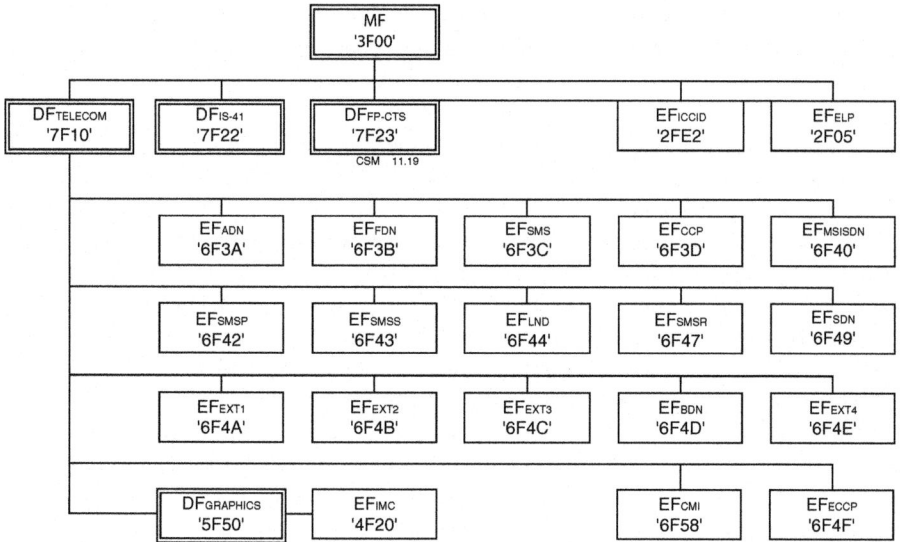

Fig. 6.3 SIM file structure (after [14])

Most of the contents found in all these files (and mainly the data that concern network use) have increased value as evidence. The fact that the user usually can't directly access them, and as such it is more difficult for him to delete them, is of great importance. A criminal, not knowing these details, will not be able to erase incriminating data leaving behind valuable evidence. Moreover, it is possible to verify these data in comparison to the ones the provider retains and spot a possible forgery.

In the next pages, we will describe some of these files and the information they hold. We are also giving the hexadecimal and the decimal value of the respective Elementary File (EF) for the interested readers who wish to experiment using the right AT commands.

6.7.2   International Mobile Subscriber Identity

For each SIM worldwide, a unique 15 digit number is assigned, named IMSI (International Mobile Subscriber Identity). IMSI number allows the SIM to be identified, even if the card itself has expired and its use is not possible in the network. This information is present in EF_{IMSI} (file identifier $6F07_h = 28423_d$)

The IMSI structure is the following:

MCC+MNC+MSIN

where:

MCC (Mobile Country Code) is the three digits country code and MNC (Mobile Network Code) is the two digits mobile telephony network code (three digits in USA and Canada). MSIN (Mobile Subscriber Identity Number) code follows. This is the user's mobile telephony serial number and consists of 10 digits (9 digits in USA and Canada). MCC numbers are described in ITU E.212 standard as described in Chap. 3. MNCs are usually serially allocated to providers.

6.7.3 Integrated Circuit Card Identifier

Even before she places the SIM in the suitable reader, the examiner can recognize the country that has issued the specific card as well as the specific network. This is possible by observing the serial number that is printed on the plastic wrapping of the card (Fig. 6.4).

The same number is also programmed in the corresponding file inside the SIM. (EF_{ICCID}, file identifier $2FE2_h = 12258_d$). It is a unique serial number (ICCID-Integrated Circuit Card Identifier) which corresponds to the printed card circuit and is based on ITU-T E.118 [36] with this form:

89(meaning telecommunications) + Country code (according to ITU-T E164) + MNC (as it was analyzed previously) + Serial number.

Country calling codes according to ITU-T E164 are the known codes we use in the phone network when we call foreign countries. For example, Greece possesses number 30, Germany possesses number 49, France 33, Italy 39, and so on. Thus for a Greek SIM card that belongs for example to WIND network, the ICCID identifier begins with 893010 as seen in Fig. 6.5 (bear in mind that digits are swapped in the output, that is, 98 is 89, 03 is 30, and so on).

Fig. 6.4 ICCID printed on the SIM

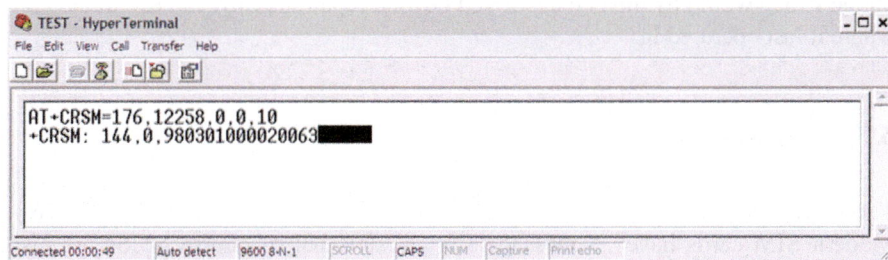

Fig. 6.5 ICCID extracted from the SIM

6.7.4 Location Information File and Broadcast Control Channel File

It is very often the case that the card arrives to the examiner in a turned off mobile phone or even without a phone at all. Apart from the originating country and network, it is also possible to pinpoint the location in which the SIM card was used for the last time.

This is possible using the Location Information file (EF_{LOCI}, file identifier $6F7E_h = 28542_d$) This file has stored the subscriber's last temporary identity (TMSI) and its last location area identifier (LAI). TMSI is a temporary identity (in contrast to IMSI which is a permanent one). It is used for security reasons since it is wise to transmit the user's permanent identity as little as possible in the network. Indeed, if IMSI was transmitted frequently, then a specific user could be specified and easier intercepted as described in Chap. 3.

The LAI code (Location Area Identification) that is also present refers to the location and has the form MCC+MNC+LAC.

As we have previously seen, MCC and MNC codes identify the country and the network. The LAC number (Location Area Code) determines a wider area which includes hundreds or even thousands of cells. Unfortunately, for the evidence examiner, the LAC area number is not referring to the actual cell, which in an urban area would determine the location in a radius of a few hundred meters or even less in dense populated areas.

However, the examiner can combine the LAC information with the BCCH (Broadcast Control Channel) list that is also stored in a respective file (EF_{BCCH}, file identifier $6F74_h = 28532_d$). The broadcast control channel information file EF_{BCCH} holds the channel number of the current (or the last, when the phone is switched off) broadcast communication control channel and its six neighboring channels. This information remains stored in the SIM even after the device's deactivation and it is renewed as the device moves. Although still missing the exact cell where the phone last functioned, the examiner can possibly correlate the BCCH channels to given cells in the specific location area.

In order to better understand the previous paragraph, let's use an example. Suppose law enforcement finds a SIM card during the baggage search of a suspect. Extracting the data as previously reported, the coarse area in which it was used the last time can be spotted. Equally impressive is the example where a mobile phone device is found in the sea. The SIM itself is waterproof and in contrast to the mobile phone which would by then be destroyed beyond repair, its SIM can reveal the area where it last operated in.

6.7.5 SMS Storage File

The SIM can also store SMS messages (EF_{SMS}, file identifier $6F3C_h = 28476_d$). Modern SIM cards have 35 message storage slots but of course SMSs are also stored in the mobile phone's memory when the card uses all the available space or if the mobile phone's manufacturer has preselected the mobile phone's memory as the main storage area. In case the message is actually stored in the SIM card, we can read the special file from the SIM file structure getting all the relevant information. The first byte of each SMS storage slot indicates the message's status and can have the following binary values and meanings:

00000000 Unused
00000001 Read incoming message
00000011 Unread incoming message
00000101 Outgoing and already sent message
00000111 Outgoing message which still has not been sent

When the user deletes a text message, instead of the message being fully erased, the only change taking place is that this first byte gets the value 0, marking the slot free for use. The remaining contents at the location that the message existed remain intact. Let's assume a message that it has already been read: its corresponding first byte has the value 1. If the message is erased, then this byte will change to 0 but the message will continue to exist, stored in the same memory slot, until it is overwritten by another message (since this memory slot has been marked as free). This way, by direct reading of this file, the forensics examiner can recover previously erased messages. In practice, certain mobile phones proceed a step further with message deletion, filling the occupied area with binary "1." This unfortunately renders the process unusable and the data retrieval impossible.

In addition to the SMS storage file, another file exists where the SMS service center parameters are stored, including the messages center number, the default validity period, the default alphabet to be used, etc. Fig. 6.6 shows an example of this file, as well as a message read. Note the first byte of the message being "01"and denoting that the message was already read.

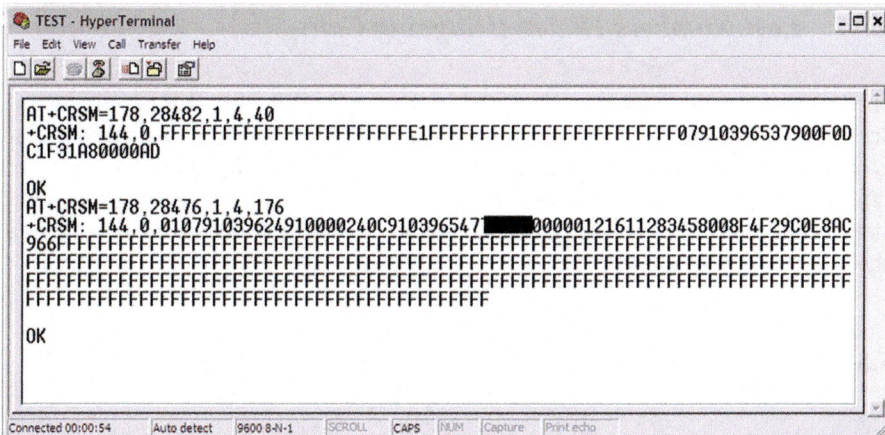

Fig. 6.6 SMS service parameters and an SMS read

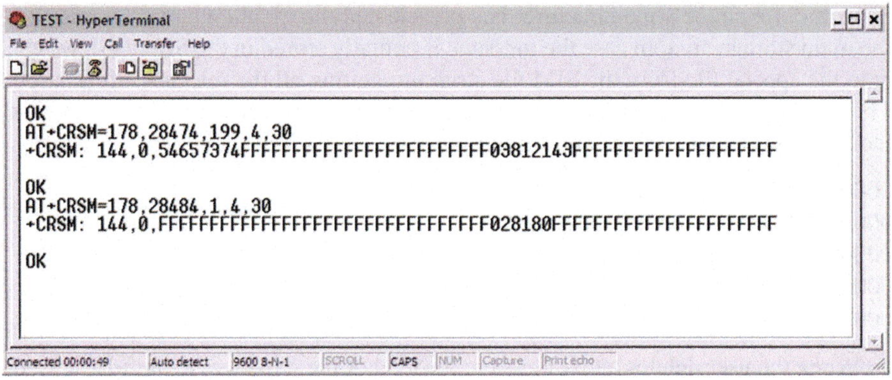
TEST - HyperTerminal
File Edit View Call Transfer Help
OK
AT+CRSM=178,28474,199,4,30
+CRSM: 144,0,54657374FFFFFFFFFFFFFFFFFFFFFFFF03812143FFFFFFFFFFFFFFFFFFFF
OK
AT+CRSM=178,28484,1,4,30
+CRSM: 144,0,FFFFFFFFFFFFFFFFFFFFFFFFFFFFFFFF028180FFFFFFFFFFFFFFFFFFFFFF
OK
Connected 00:00:49
Auto detect
9600 8-N-1
SCROLL
CAPS
NUM
Capture
Print echo

Fig. 6.7 Store contact and last number dialed

6.7.6 *Abbreviated Dialing Numbers (Contact List)*

The contacts catalog along with the stored messages forms one of the most important pieces of evidence. For the contact list storage file (EF_{ADN}, file identifier $6F3A_h = 28474_d$), older SIM cards offered 100 storage slots while the newer ones offer 250 positions. When the user erases a contact from the list almost always, the contents of the corresponding storage slot are overwritten with binary "1." In contrary to the message deletion that we just described, previous contacts can't be recovered. On the other hand, the memory slots in this file are serially assigned; therefore, it is possible to extract information about whether some contact number has been erased. If for example an empty slot is found between slot 34 and slot 36, we can conclude that a contact was also stored in place 35 before being deleted. Fig. 6.7 shows an example of stored contacts.

6.7.7 *Last Numbers Dialed (Outgoing Calls)*

The last ten outgoing calls can be stored in the relative SIM file (EF_{LND}, file identifier $6F44_h = 28484_d$). Most manufacturers, however, do not use it because they prefer to use mobile phone's memory for the same functionality. When they do use it, the data stored look like that presented in Fig. 6.7. It is interesting to note at this point that the SIM does not store incoming call details. They are found only in the mobile phone's memory as we will see below.

6.8 Device Data

Despite the impressive capabilities of the SIM card, it is essential for certain kind of data to be stored in the phone's memory [37]. Two memory types are used, the NAND-flash and the NOR-flash. The actual data stored in the internal Flash memory depend on the manufacturer. They usually include the IMEI, time settings, sound settings, volume settings, SMS messages, calendar-alarm clock, missed and answered calls, contacts catalog with multiple fields (in contrast to SIM's catalog that offers only one telephone field per contact), and so on. Executable files and applications or games and multimedia files such as pictures, video, and sound recordings are also stored in the phone's internal or external memory. Adding internet browsing capability, the information stored in mobile phones extends and includes lists of web pages that were visited by the user, favorite web pages, Wi-Fi access points, and so on.

Beyond the current data, older or even deleted data can be found in the "depths" of the mobile phone's memory. They include text messages, pictures, MMS, simple notes and calendar notes, contacts, and so on. Such information is conditionally possible to be (partially or totally) recovered after deletion. Even data regarding SIM cards that have been used with the phone can have been stored in memory (e.g. previous card's IMSI that was used in the same mobile)

The examiner is called to extract these data without altering them [38, 39]. For this purpose, he can resort to the memory dump byte to byte for NOR memory (or page to page for NAND type memory). This is achieved using special software and hardware tools that "clone" the memory contents [40]. In contrast to the SIM card, this way she can receive a complete data image in physical layer. The whole process is particularly complicated from technical point of view because the data are unstructured, and they have to be translated in a specific file system.

The basic functionality of tools used to dump memory contents is to upload new firmware versions, to upgrade, repair, and debug mobile phones. They can be officially provided by the manufacturers. Unofficially, third party companies and private individuals are selling relevant tools. The use of unofficial tools is sometimes related to various illegal activities such as unlocking and serial numbers changing.

The precious contribution of these tools in digital evidence extraction lies in the fact that they can operate even when the mobile phone is turned off, locked, blocked, (partially) broken, and so on. Naturally, such powerful tools are dangerous. They can alter data if the user is not careful and experienced enough thus destroying whatever evidence existed in first place.

There is no standardized way or a common contact for the connection (each manufacturer uses a different one even between his own models) while it is often required to access special internal contacts, found in the printed circuit board of the phone. One of these contacts is JTAG (Joint Test Action Group) described in IEEE 1149.1 standard [41]. The functionality offered is a very powerful feature. It is a special debug interface initially used to test the printed circuit boards of electronic devices. It further evolved to allow checking, monitoring, and debugging of embedded systems. For forensics use, it can undertake the memory dump, but it is difficult to find the specific test points, and usually documentation is proprietary information not readily available. A typical JTAG connection point and a cable (from a different mobile phone than the one depicted) are seen in Figs. 6.8 and 6.9.

Despite the advantages of the memory dump method, certain problems exist. The basic problem is that there is no way to detect if external changes have taken place in the flash memory (therefore the data may have been modified). At the same

Fig. 6.8 JTAG pinouts

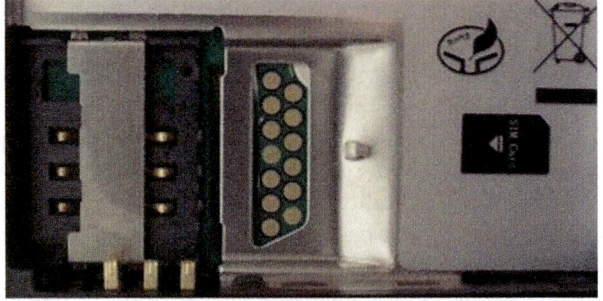

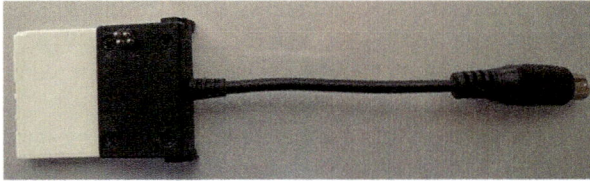

Fig. 6.9 JTAG cable

time, mobile phones use memory managers which dynamically distribute and reshape the memory data. This is done in order to achieve optimal memory use and minimal wear leveling. Practically, this means that the precise location where a specific piece of information/evidence resides in the memory dump changes each time the dump is performed! In addition, searching for valuable information in a dump of some Gigabytes in size is a really hard and time-consuming work, while, on top of that, some phones have encrypted memory contents. Moreover, we should note that certain mobile phones delete part of the data stored in their memory when a different SIM is inserted other than the one that was lastly used. Apart from hardware tools, various software data suites exist that allow high level access to the memory (using the mobile phone's operating system). As such the interaction with the phone is increased; this is why they have to be certified for forensics use.

6.9   External Memory Dump

If no other method can be used (e.g. because the mobile phone is partially or totally destroyed), it is also possible to detach/desolder the integrated memory circuits using special precision SMD (surface mount device) soldering/desoldering stations. Following that, external memory dump takes place using the right hardware tools. This process guarantees that no data "infection" happens since the mobile phone remains switched off. It faces, however, the serious danger of the complete circuit destruction during that delicate process of detachment/desoldering. Moreover, the mobile phone should be disassembled in order to extract the integrated memory circuit. Considering these difficulties, this method is the least preferred and the one that the analyst will ultimately resort to if nothing else seems to be working.

6.10   External Memory Cards and Computers

Given the increased multimedia capabilities of modern phones, external memory cards have become particularly widespread. They provide large data storage capability for videos, photographs, music, and other file types. At the same time, they can be used in computers and other devices such as PDAs, MP3 players, cameras, and video cameras, being a perfectly usable data transfer medium. A series of different interfaces and standards exist for their interconnection. Luckily, these interfaces and their internal workings do not reach the complexity of the respective ones found in mobile phones. Indeed, it is relatively easy to read those memory types because there is a wealth of products available, covering beginners as well as advanced users. Once again, the principles of extracting data from the memory apply, as analyzed previously.

It should be noted that these cards can be transferred to the computer, or respectively the computer can be connected to the mobile phone (e.g. for backup copies). Consequently, not only the mobile phone but also the hard disk and the computer memory might contain data from possible inter-connection of the computer with the mobile phone. This means that the examiner's work is not limited to the mobile phone but it further continues to the computer. We won't extend further our analysis because digital evidence extraction from computers is a whole subject by itself.

6.11   Evidence in the Operator's Network

Apart from data that the expert can extract from the mobile phone, the operator maintains logs with the calls, SMS exchanged, data use, geographical data, and many more. The Home Location Register (HLR) which is one of the most important elements of mobile phone networks keeps details about the user, identification numbers such as IMSI and the SIM serial number, the PIN/PUK codes, the services he has subscribed to, and so on.

Of particular importance are the Call Detail Records (CDR). They have analytical information about the calling and called number, the date, duration, the cells that have served the call, and so on. This information can be also correlated to call logs evidence extracted from the phone. It is usually the first thing law authorities ask from the operator to provide, following a warrant. At this point, different retention laws mandate the duration that the provider has to keep these data.

6.12   Conclusion

The everlasting increase in memory capacity combined with smaller dimensions and form factors allows the reception, storage, and processing of more and more data in portable devices and mobile phones. In case their owner is involved in criminal/illegal activities, these data can prove evidence and can play an important role for the law enforcement and judicial authorities.

Specifically for mobile phones, the digital forensics analysis is a permanent and difficult fight against criminals. In contrast to computer forensics, mobile forensics take place on the original data (in case of the SIM). This makes the task of the analyst considerably harder. At the same time, thousands of different models exist, with different characteristics and interfaces. A relative lack of forensics standards and exceptional complexity in the providers' networks make things worse, while the examiner is often called to work under pressure and narrow time limits.

Beyond the technical details and complexity, nontechnical issues play an equally important role. The digital forensics examiner should be constantly up to date with the field's technological developments and follow specific and established procedures. Once again, education is what will help combine the technical with the procedural part so that the work of the analyst can substantially help the proper trial of the case.

The volume of evidence stored in mobile phones and SIM cards is multiplied in a fast pace. Experimental SIM cards with embedded Wi-Fi transceivers [42] as well as SIM cards with embedded GPS receivers [43] are already being tested in the industry. It is almost certain that in the future, such data evidence will be present even in embedded systems found in almost every device we use in our daily life. Since modern criminals will be basing their malicious actions on technology, digital evidence analysis is more necessary than ever. This is why, in this chapter, we attempted to describe the methodology and procedures encountered in collecting and analyzing evidence from mobile phones.

References

1. Harrill DC, Mislan RP (2007) A small scale digital device forensics ontology. Small Scale Digital Device Forensics Journal 1:1
2. Richard P. Mislan, Cellphone crime solvers, IEEE Spectrum, 07/2010
3. Androulidakis I (2009) Digital evidence in mobile phones, IT security professional magazine. Issue 13:36–39
4. Gratzer V, Naccache D (2006) Cryptography, law enforcement, and mobile communications. IEEE Security Privacy 4(6):67–70
5. AlZarouni M (2006) Mobile handset forensic evidence: a challenge for law enforcement. School of Computer and Information Science, Edith Cowan University, Mount Lawley
6. International Organization on Computer Evidence Guidelines for Best Practice in the Forensic Examination of Digital Technology http://www.ioce.org/core.php?ID=5
7. ACPO Good Practice Guide for Computer-Based Evidence. ACPO,http://www.7safe.com/electronic_evidence/ACPO_guidelines_computer_evidence_v4_web.pdf
8. Jansen W (2007) Guidelines on cell phone forensics. NIST SP 800–101
9. Jansen W (2004) Guidelines on PDA Forensics. NIST SP 800–72
10. McCarthy P (2005) Forensic analysis of mobile phones. School of Computer and Information Science, University of South Australia, Mount Lawley
11. Gary C (2007) Kessler. Cell Phone Analysis. In: DoD Cyber Crime Conference
12. ISO/IEC 7816-1 (1998) Identification cards—Integrated circuit(s) cards with contacts—Part 1: Physical characteristics
13. ETS 300 977 (GSM 11.11 version 5.5.0), European Telecommunications Standards Institute, Digital cellular telecommunications system (Phase2+); Specification of the Subscriber Identity Module Mobile Equipment (SIM ME) interface, May 1997

14. 3GPP TS 51.011 (2005) Technical Specification Group Core Network and Terminals; Specification of the Subscriber Identity Module -Mobile Equipment (SIM—ME) interface
15. Quantaq Solutions (2006) USIMcommander
16. Ayers R, Tools CPF (2007) An overview and analysis update. NIST IR 7387
17. Ayers R (2006) An overview of cell phone forensic tools. NIST
18. Mokhonoana PM, Olivier MS (2007) Acquisition of a Symbian Smart phone's content with an on-phone forensic tool, SATNAC2007
19. Jansen WA, Delaitre A (2007) Reference material for assessing forensic SIM tools. Paper No ICCST 2007–74
20. Jansen WA, Delaitre (2006) A overcoming impediments to cell phone forensics. NIST
21. Casadei F (2006) Forensics and SIM cards: an overview. Int J Digital Evidence Fall 5(1):2006
22. Vedder K, Cards S (2006) ETSI Future Security Workshop: the risks, threats and opportunities
23. Vedder K, Cards S (2007) 2nd ETSI security workshop: future security
24. Keith E (2008) Mayes and Konstantinos Markantonakis. Smart cards, tokens, security and applications. Springer, Heidelberg
25. Richter T (2002) Chipkarten im Mobilfunk
26. Tual J-P (2004) Introduction à la carte à puce, Axalto
27. Kanninen A (2000) Emerging problems with smart card technologie
28. Posegga J (2003) Smartcards. UKA SS
29. Rankl W, Effing W (2004) Smart cards in telecommunications, in smart card handbook, 3rd edn. Wiley, New York
30. Rankl W (2004) Protocol of the communication between SIM and mobile phone
31. Rankl W (2004) Smart card training, some basics about the SIM
32. Guo H. Smart cards and their operating systems. HUT, Telecommunications Software and Multimedia Laboratory
33. ETSI TS 101 220 (2007) Smart Cards; ETSI numbering system for telecommunication application providers
34. ETSI TS 102 222 (2003) Integrated circuit cards (ICC); Administrative commands for telecommunications applications
35. Witteman M (2001) Everything you always wanted to know about Smart Cards Black Hat
36. ITU, ITU-T (2006) Recommendation E.118, The international telecommunication charge card
37. Androulidakis I (2010) The mobile phone has a memory and remembers. IT Security Professional Magazine 14
38. Willassen SY (2005) Forensic analysis of mobile phone internal memory advances in digital forensics. IFIP WG 11.9 conference on Digital Forensics
39. Breeuwsma M, de Jongh M, Klaver C, van der Knijff R, Roeloffs M (2007) Forensic data recovery from flash memory. Small Scale Digital Device Forensics J 1(1)
40. Al-Zarouni M (2007) Introduction to mobile phone flasher devices and considerations for their use in mobile phone forensics. In: Proceedings of The 5th Australian digital forensics conference
41. IEEE, IEEE 1149.1 (2001) IEEE standard test access port and boundary scan architecture
42. Middleton J (2010) Sagem embeds wifi in SIM. http://www.telecoms.com/18192/sagem-embeds-wifi-in-sim
43. James Middleton, Estonian Operator Trials SIM-Embedded GPS tech, http://www.telecoms.com/17811/estonian-operator-trials-sim-embedded-gps-tech,2010

Chapter 7
Conclusion

Abstract This chapter concludes the book presenting a short overview of the situation. Hopefully, with the presentation of vulnerabilities and the way malicious users take advantage of them, the book has helped raise user awareness. Since GSM is not considered anymore, secure technology users should always use common sense and the necessary precautions to enjoy mobile phones securely. To help in that, we are giving in a condensed list as much as possible practical security advices.

Keywords: Mobile phone security • Security practices • Practical security • Mobile phone confidentiality • Mobile phone integrity • Mobile phone availability • GSM security • User awareness

7.1 In General

In this book, we tried to raise user awareness in regard to security and privacy threats present in the use of mobile phones. We focused on practical issues, skipping theoretical analysis of algorithms and standards. As is the case with every modern technology, threats and fraud do exist and the user should be vigilant. Mobile phones can be used for criminal actions and are a valuable source of evidence for solving cases. Being almost dependent on mobile phones, availability issues can cause a great deal of inconvenience. Even worse, important privacy intrusions include not only the interception of voice, SMS, and data but also the location of the user. The mobile phone is indeed a live transmitter which we carry on the whole day, revealing our whereabouts. Finally, fraudsters can target our mobile phone. This gets more and more possible given the increased use of mobile phones to surf the Internet enjoying services such as mobile commerce.

GSM used to be a relatively secure standard. This is clearly not the case anymore. Some of its security shortcomings were the following: "closed" design (security through obscurity), unsecure core network, bad implementations, lack of mutual authentication, and internal fraud issues. With principles such as public design, mutual authentication, better encryption algorithms with longer keys, encryption in the core network, and other solutions yet to be implemented, 3G and future systems are more secure than GSM.

© Springer International Publishing Switzerland 2016   111
I.I. Androulidakis, *Mobile Phone Security and Forensics*,
DOI 10.1007/978-3-319-29742-2_7

Even then, attackers will eventually find new ways of abusing the systems, while the research community will continuously actively challenge their security. As such, users should always use common sense and the necessary precautions to enjoy mobile phones securely. Below we are giving in a condensed list as much as possible practical security advices.

7.2   Practical Security Advices

- Enable the PIN in your SIM card
- Keep your PIN and other security codes secret
- Use a password in the screen saver
- Do not lend your phone, not even for a few minutes
- Do not leave your phone unattended
- If you are obliged to hand over your phone, in places where it is not allowed to carry it on, place it in a tamper-proof bag/envelope
- Personalize your phone so it can't be swapped with another one
- Do not carry your phone in critical meetings or remove its battery
- Write down the IMEI of your phone
- Do not save sensitive data
- Keep firmware and operating system updated
- Perform backups frequently
- Familiarize yourself with the screen's indicators/icons and pay attention to them
- Enable netmonitor to check the encryption level
- Be suspicious if screen light suddenly comes up
- Familiarize yourself with call waiting and conference call announcement tones (beeps heard periodically during the call)
- Disable GPS if not needed
- Do not visit insecure web pages using the mobile
- Do not accept unknown files through Bluetooth, WAP, email, and MMS
- Do not accept any unsolicited connections (e.g. via Bluetooth)
- Do not install unknown applications
- Check your bills looking for calls and SMSs to unknown numbers, as well as for data usage
- Use an antivirus if exists for your phone
- Consider the use of cryptophones or software suites to encrypt communication
- Consider applications that encrypt stored data
- Consider applications that allow the remote deletion of data in case the phone is stolen. Such applications can possibly block the phone too
- Do not blindly trust the Caller ID or the originator of an SMS
- Keep Bluetooth switched off (or at least invisible)
- Use a lengthy PIN in every Bluetooth pairing
- Do not pair devices using Bluetooth in unsecure areas
- Check periodically the Bluetooth trusted devices list

- Enable encryption during Bluetooth pairing with PC
- If under any kind of Bluetooth harassment attack, move away
- Perform precautionary Bluetooth searches to locate devices you have forgotten to switch off or hostile ones
- Surges in consumption leading to drastically reduced battery life could be a hint of malicious software running
- Continuous (not periodic every few hours) humming sound (the same you hear when you speak in the mobile phone nearby a loudspeaker) while you do not use the phone should be examined
- Check periodically the internet access points registered
- Check running processes
- Do not enter your mobile phone number in various web pages that request it (free games, ringtones, logos, etc.)

About the Author

Dr. Iosif I. Androulidakis has an active presence in the ICT security field. He has authored more than 90 publications (including six books) and has presented more than 120 talks and lectures in international conferences and seminars in 20 countries while he has participated in many security projects.

Holding two Ph.D.s his research interests focus on security issues in telecommunication systems where he has more than 20 years of experience. Part of his research has led to the granting of five patents. During his career, he collaborated with Telecom Operators, National Police Cybercrime departments in many countries, the European Police Academy (CEPOL), the European Public Law Center, the Southeastern Europe Telecommunications & Informatics Research Institute, universities and research centers, vocational training institutes, and the media and private security consulting firms.

Dr. Androulidakis has also acted as a reviewer in an extended array of scientific conferences and journals, as a Programme Committee member in 26 conferences and as a chairman in 8 conference sessions. Finally, he is a certified ISO9001 (Quality Management System) and ISO27001 (Information Security Management System) auditor and consultant who has an active presence in the ICT security field having authored more than 50 papers and having presented more than 100 talks and lectures in international conferences and seminars in 18 countries. Holding a Ph.D. in electronics, his research interests focus on security issues in PBXs (private telephony exchanges) where he had 18 years of experience, as well as in mobile phones and embedded systems. Part of his research has led to the granting of two patents. During his career, he collaborated with telecom operators, national police cybercrime departments, the European Police Academy (CEPOL), the European Public Law Center, universities and research centers, vocational training institutes, and the media and private security consulting firms. Mr. Androulidakis is also a member of IEEE (Technical Committee on Security and Privacy) and ACM (Special Interest Group on Security, Audit and Control). Finally, he is a certified ISO9001:2000 (Quality Management System) and ISO27001:2005 (Information Security Management System) auditor and consultant.

© Springer International Publishing Switzerland 2016   115
I.I. Androulidakis, *Mobile Phone Security and Forensics*,
DOI 10.1007/978-3-319-29742-2

Index

© Springer International Publishing Switzerland 2016

I.I. Androulidakis, *Mobile Phone Security and Forensics*,
DOI 10.1007/978-3-319-29742-2

Printed by Printforce, the Netherlands